Lecture Notes in Networks and Systems

Volume 92

The series "Lecture Notes in Networks and Systems" publishes the latest developments in Networks and Systems—quickly, informally and with high quality. Original research reported in proceedings and post-proceedings represents the core of LNNS.

Volumes published in LNNS embrace all aspects and subfields of, as well as new challenges in, Networks and Systems.

The series contains proceedings and edited volumes in systems and networks, spanning the areas of Cyber-Physical Systems, Autonomous Systems, Sensor Networks, Control Systems, Energy Systems, Automotive Systems, Biological Systems, Vehicular Networking and Connected Vehicles, Aerospace Systems, Automation, Manufacturing, Smart Grids, Nonlinear Systems, Power Systems, Robotics, Social Systems, Economic Systems and other. Of particular value to both the contributors and the readership are the short publication timeframe and the world-wide distribution and exposure which enable both a wide and rapid dissemination of research output.

The series covers the theory, applications, and perspectives on the state of the art and future developments relevant to systems and networks, decision making, control, complex processes and related areas, as embedded in the fields of interdisciplinary and applied sciences, engineering, computer science, physics, economics, social, and life sciences, as well as the paradigms and methodologies behind them.

** Indexing: The books of this series are submitted to ISI Proceedings, SCOPUS, Google Scholar and Springerlink **

More information about this series at http://www.springer.com/series/15179

Mostafa Ezziyyani

Editor

Advanced Intelligent Systems for Sustainable Development (AI2SD'2019)

Volume 6 - Advanced Intelligent Systems for Networks and Systems

 Springer

Editor
Mostafa Ezziyyani
Faculty of Sciences
and Techniques of Tangier
Abdelmalek Essaâdi University
Tangier, Morocco

ISSN 2367-3370   ISSN 2367-3389   (electronic)
Lecture Notes in Networks and Systems
ISBN 978-3-030-33102-3   ISBN 978-3-030-33103-0   (eBook)
https://doi.org/10.1007/978-3-030-33103-0

This Springer imprint is published by the registered company Springer Nature Switzerland AG
The registered company address is: Gewerbestrasse 11, 6330 Cham, Switzerland

Preface

After the success of the first edition of AI2SD'18, this second International Conference on Advanced Intelligent Systems for Sustainable Development 2019 (AI2SD'2019) held in the beautiful Marrakech city, Morocco, continues to establish new scientific and professional bridges in the fields of advanced intelligent systems and technology. This event has distributed multidisciplinary technical programme in "**Digital Transformation is lever of the Industrial Revolution 4.0**" that draws attention and participation from over 400 researchers in the field of advanced intelligent systems for sustainable development. It is also expected a great opportunity among students, scientists, researchers, and industrial sectors to exchange their ideas with others in the knowledge of sustainable development and establish an excellent network with a new colleague.

At this special event, 15 international distinguished keynote speakers and a large number of invited speakers will deliver their outstanding research works in various fields of advanced intelligent systems joining together with 253 oral presentations and about 62 poster presentations that would be a great opportunity to all participants to share their recent research information and knowledge among each other. In addition to that, a free International Summer School for Big Data and Data Mining Models will take place in our conference, 8 courses animated by 8 international professors for the benefit of 100 PhD students.

All papers in this issue have been subjected to a peer-reviewed process based on their originality and quality. The aim of the conference is to be a platform for researchers to present their latest research works and to exchange their ideas. The topics covered in this conference have covered the following aspects:

- Advanced Intelligent Systems for Education and Intelligent Learning System
- Advanced Intelligent Systems for Sustainable Development Applied to Industry
- Advanced Intelligent Systems for Sustainable Development Applied to Economy
- Advanced Intelligent Systems for Multimedia Processing and Mathematical Modelling
- Advanced Intelligent Systems for Information and System Security

- Advanced Intelligent Systems for Supporting Decision and Prediction
- Advanced Intelligent Systems for Sustainable Development Applied to Agriculture
- Advanced Intelligent Systems for Sustainable Development Applied to Energy
- Advanced Intelligent Systems for Sustainable Development Applied to Health
- Advanced Intelligent Systems for Sustainable Development Applied to Environment
- Advanced Intelligent Systems for Computational Web and Big Data Analytics
- Advanced Intelligent Systems for Networking Systems and IoT
- Advanced Intelligent Systems for GIS and Spatial Data
- Advanced Intelligent Systems for Electrical Engineering

Overview About AI2SD Editions

The International Conference on Advanced Intelligent Systems for Sustainable Development applied to Agriculture, Energy, Health, Environment, Industry and Economy organized each year is one of the best international amalgamations of eminent researchers, students, and delegates from both academia and industry where the collaborators have interactive access to emerging technology and approaches globally. The conference focus is to have International Scientific Panel Discussion on Advanced Technologies and Intelligent Systems for Sustainable Development Applied to Education, Agriculture, Energy, Health, Environment, Industry and Economy. For each edition, AI2SD brings the bright minds to give talks that are idea-focused on a wide range of subjects, to foster learning and inspiration and provoke conversations with attractive opportunities leading to competitive advantages. AI2SD is your best opportunity to influence the principal participants from the agricultural energy, health, and environment community. Additionally, researchers working in those fields will have an access to precise method for propelling their research. Also, AI2SD editions admit to sharing of best works by means of personal talks, sharing knowledge, and experience. AI2SD editions offer oral, poster sessions, tutorials, training, and professional meetings. The programme of the AI2SD Conference intends to foster interaction so as to open the way to future cooperation between participants. The submitted papers are expected to cover the state-of-the-art technologies, theoretical concepts, standards, product implementation, ongoing research projects, and innovative applications of sustainable development in agriculture, energy health, environment, industry, and economy sensing. The fundamental and specific topics of AI2SD are big data analytics, wireless sensors, IoT, geospatial technology, engineering and mechanization, modelling tools, risk analytics, preventive systems.

Foreword

It is with deep pleasure and satisfaction that I write this Foreword to the Proceedings of the International Conference on Advanced Intelligent Systems for Sustainable Development (AI2SD'19) held in the wonderful Moroccan city of Marrakech in 8–11 July 2019. This year, AI2SD'19 projected into broader hot research topics that strives to stimulate study and research in favour of socio-economic sustainable development.

AI2SD'19 consists of technical, invited sessions, and keynote and tutorial sessions covering the state-of-the-art and advanced research work on intelligent systems applied to agriculture, environment, health, energy, economy, and industry along with themes related to big data, networking, computer vision, natural language processing, and other scopes. The papers contributed with the most recent scientific knowledge known in the aforementioned. The Technical Program committee (TPC) will include more than 300 of them in these volume proceedings, given their originality and relevance to the conference scopes. TPC will also include 12–14 keynote speeches addressing hot topics related to the conference themes. The papers accepted and presented in AI2SD'19 will be published in proceedings as special issue of Springer proceedings books within the Advances in Intelligent Systems and Computing Series/Lecture Notes in Networks and Systems/Lecture Notes in Electrical Engineering (into seven volumes). Moreover, a number of selected high-impact full-text papers will also be considered for the special journal issues as extended version. All our thanks and greetings are addressed to the committee chairs for their great works, in either the organization or the review process.

Acknowledgements to AI2SD'2019 participants

We would like to express our heartfelt thanks to each of you who participated at AI2SD'2019 Conference in Marrakesh. We have had four very rewarding days, filled with interesting keynote lectures, sessions, poster and demo presentations, summer school as well as a wide selection of assistive technology. We have had many interesting discussions with people all over the world, and we hope you have as well. Hopefully, we hope that all participates enjoyed both the scientific part programmes and that used the opportunity to extend their existing networks. I am sure that the cooperation with most of you will continue in the near future. We would like to give special thanks to the speakers of the conference and all partner institutes supporting us during the event, and furthermore to effective teams who did outstanding work in organizing the event.

Mostafa Ezziyyani
AI2SD'2019 General Chair

Organization

General Chair

Mostafa Ezziyyani   Faculty of Sciences and Technologies of Tangier, Computer Sciences Department

Co-chairs

Hajar Mousannif   Faculty of Semlalia Cadi Ayyad University, Marakkech, Morocco

Loubna Cherrat   National School of Commerce and Management, Abdelmalek Essaâdi University, Tangier, Morocco

Keynotes Speakers

Ikken Badr   IRESEN, Rabat, Morocco

Khalil Ismail   Institute of Telecooperation, Johannes Kepler University Linz, Austria

Lloret Mauri Jaime   Polytechnic University of Valencia, Spain

Kacprzyk Janusz   Polish Academy of Sciences, Poland

Mohammed Seaid   Durham University, UK

Lorenz Pascal   University of Nancy, France

Alexander Gelbukh   Instituto Politécnico Nacional, Mexico

Kamal Labbassi   Chouaib Doukkali University, Morocco

Mohammed Ezziyyani   Faculté Polydisciplinaire de Larache, Morocco

A. Samed Bernoussi   Faculty of Sciences and Technologies of Tangier, Morocco

M. F. Abdel Wahab   BU, Egypt

Kettani Nasser   KDC, Casablanca, Morocco

Khalid Zine-Dine   ChDU, FS El Jadida, Morocco

Ouazar Driss   M.VI.P U, Marrakech, Morocco

Hammari Jalal   Engagement Manager, Industry 4.0 Specialist at Siemens, Germany

Khadija Sabiri   JCI Casablanca, Morocco

TPC Chairs

Pascal Lorenz, France
Puerto Molina, Spain
Jaime Lloret Mauri, Spain
Joel Rodrigues, Portugal
Mohammed Bahaj, Morocco
Mohammed Ezziyyani, Morocco
Abdel Ghani Laamyem, Morocco

Khalid Zine-Dine, Morocco
Ahmed Azouaoui, Morocco
Philippe Roose, France
El Metoui Mustapha, Morocco
Jbilou Mohammed, Morocco
Issam Qaffou, Morocco
Mustapha Zbakh, Morocco

PhD Organizing Committee

Maroi Tsouli Fathi
Marwa Zaryouli
Sara Khrouch
Yassine Doukali Fenni

Soumaya El Mamoune
Mohammed Rida Ech-Charrat
Mohammed Redouane Taghouti

Local Organizing Committee

Jihad Zahir, FSSM
Issam, Qaffou, FSSM
El Bachari Essaid, FSSM

El Adnani Mohamed, FSSM
Hakim El Boustani, FSSM
Agouti Tarik, FSSM

Scientific Committee

Pedro Mauri, Spain
Sandra Sendra, Spain
Lorena Parra, Spain
Oscar Romero, Spain
Kayhan Ghafoor, China
Jaime Lloret Mauri, Spain
Yue Gao, UK
Faiez Gargouri, Tunis
Mohamed Turki, Tunis
Abdelkader Adla, Algeria
Souad Taleb Zouggar, Algeria
Bakhta Nachet, Algeria
Danda B. Rawat, USA
Tayeb Lemlouma, France
Mohcine Bennani Mechita, Morocco
Tayeb Sadiki, Morocco

Mhamed El Merzguioui, Morocco
Abdelwahed Al Hassan, Morocco
Mohammed Boulmalf, Morocco
Abdellah Azmani, Morocco
Kamal Labbassi, Morocco
Jamal El Kafi, Morocco
El Hassan Abdelwahed, Morocco
Mohamed Chabbi, Morocco
Mohamed Riduan Abid, Morocco
Jbilou Mohammed, Morocco
Salima Bourougaa-Tria, Algeria
Ahlem Hamdache, Morocco
Mohammed Reda Britel, Morocco
Youness Tabii, Morocco
Mohamed El Brak, Morocco
Hamid Harroud, Morocco

Joel Rodrigues, Portugal
Ridda Laaouar, Algeria
Mustapha El Jarroudi, Morocco
Abdelouahid Lyhyaoui, Morocco
Ziani Ahmed, Morocco
Karim El Aarim, Morocco
Mustapha Maatouk, Morocco
Abdel Ghani Laamyem, Morocco
Abdessamad Bernoussi, Morocco
Loubna Cherrat, Morocco
Ahlem Hamdache, Morocco
Mohammed Haqiq, Morocco
Abdeljabbar Cherkaoui, Morocco
Rafik Bouaziz, Tunis
Hanae El Kalkha, Morocco
Abdelaaziz El Hibaoui, Morocco
Othma Chakkor, Morocco
Abdelali Astito, Morocco
Mohamed Amine Boudia, Algeria
Mebarka Yahlali, Algeria
Hasna Bouazza, Algeria
Zakaria Bendaoud, Algeria
Driss Sarsri, Morocco
Muhannad Quwaider, India
Mohamed El Harzli, Morocco
Wafae Baida, Morocco
Mohammed Ezziyyani, Morocco
Xindong Wu, China
Sanae Khali Issa, Morocco
Monir Azmani, Morocco
El Metoui Mustapha, Morocco
Mustapha Zbakh, Morocco
Hajar Mousannif, Morocco
Mohammad Essaaidi, Morocco
Amal Maurady, Morocco
Ouardouz Mustapha, Morocco

Mustapha El Metoui, Morocco
Said Ouatik El Alaoui, Morocco
Nfaoui El Habib, Morocco
Aouni Abdessamad, Morocco
Ammari Mohammed, Morocco
Ben Allal Laila, Morocco
El Afia Abdelatif, Morocco
Noureddine Ennahnahi, Morocco
Ahachad Mohammed, Morocco
Abdessadek Aaroud, Morocco
Mohammed Said Riffi, Morocco
Kazar Okba, Algeria
Omar Akourri, Morocco
Mohammed Bahaj, Morocco
Feddoul Khoukhi, Morocco
Pascal Lorenz, France
Puerto Molina, Spain
Herminia Maria, Spain
Abderrahim Abenihssane, Morocco
Abdelmajid Moutaouakkil, Morocco
Silkan, Morocco
Khalid El Asnaoui, France
Salwa Belaqziz, Morocco
Khalid Zine-Dine, Morocco
Mounia Ajdour, Morocco
Essaid Elbachari, Morocco
Mahmoud Nassar, Morocco
Khalid Amechnoue, Morocco
Hassan Samadi, Morocco
Abdelwahid Yahyaoui, Morocco
Hassan Badir, Morocco
Ezzine Abdelhak, Morocco
Mohammed Ghailan, Morocco
Mohamed Ouzzif, Morocco
Mohammed A. Moammed Ali, Sudan

Contents

About the Editor

Prof. Dr. Mostafa Ezziyyani IEEE and ASTF Member, received the "Licence en Informatique" degree, the "Diplôme de Cycle Supérieur en Informatique" degree, and the PhD "Doctorat (1)" degree in information system engineering, respectively, in 1994, 1996, and 1999, from Mohammed V University in Rabat, Morocco. Also, he received the second PhD degree "Doctorat (2)" in 2006, from Abdelmalek Essaadi University in distributed systems and Web technologies. In 2008, he received a Researcher Professor Ability Grade. In 2015, he received a PES Grade—the highest degree at Morocco University. Now he is Professor of Computer Engineering and Information System in Faculty of Science and Technologies of Abdelmalek Essaadi University since 1996.

His research activities focus on the modelling databases and integration of heterogeneous and distributed systems (with the various developments to the big data, data sciences, data analytics, system decision support, knowledge management, object DB, active DB, multi-system agents, distributed systems, and mediation). This research is at the crossroads of databases, artificial intelligence, software engineering, and programming.

Professor at Computer Science Department, Member MA Laboratory, and responsible of the research direction information systems and technologies, he formed a research team that works around this theme and more particularly in the area of integration of heterogeneous systems and decision support systems using WSN as technology for communication.

He received the first WSIS Prize 2018 for the Category C7: ICT applications: AQ1 E-environment, first prize: MtG—ICC in the regional contest IEEE - London UK Project: "World Talk"; the qualification to the final (Teachers-Researchers Category): Business Plan Challenger 2015, EVARECH UAE Morocco; project: «Lavabo Intégré avec Robinet à Circuit Intelligent pour la préservation de l'eau», first prize: Intel

Business, Challenge Middle East and North Africa—IBC-MENA; project: «Système Intelligent Préventif Pour le Contrôle et le Suivie en temps réel des Plantes Médicinale En cours de Croissance (PCS: Plants Control System)»; best paper: International Conference on Software Engineering and New Technologies ICSENT'2012, Hammamet, Tunis; and paper: «Disaster Emergency System Application Case Study: Flood Disaster».

He has authored three patents: (1) device and learning process of orchestra conducting (e-Orchestra), (2) built-in washbasin with intelligent circuit tap for water preservation, and (LIRCI) (3) device and method for assisting the driving of vehicles for individuals with hearing loss. He is Editor and Coordinator of several projects with Ministry of Higher Education and Scientific Research and others as international project; he has been involved in several collaborative research projects in the context of ERANETMED3/PRIMA/H2020/FP7 framework programmes including project management activities in the topic modelling of distributed information systems reseed to environment, health, energy, and agriculture. The first project aims to propose an adaptive system for flood evacuation. This system gives the best decisions to be taken in this emergency situation to minimize damages. The second project aims to develop a research dynamic process of the itinerary in an event graph for blind and partially signet users. Moreover, he has been Principal Investigator and Project Manager for several research projects dealing with several topics concerned with his research interests mentioned above.

He was Invited Professor for several countries in the world (France, Spain, Belgium, Holland, USA, and Tunisia). He is Member of USA-J1 Programme for TCI Morocco Delegation in 2007. He creates strong collaborations with research centres in databases and telecommunications for students' exchange: LIP6, Valencia, Poitiers, Boston, Houston, China.

He is Author of more than 100 papers which appeared in refereed specialized journals and symposia. He was also Editor of the book "New Trends in Biomedical Engineering", AEU Publications, 2004. He was Member of the Organizing and the Scientific Committees of several symposia and conferences dealing with topics related to computer sciences, distributed databases, and Web technology. He has been actively involved in the research community by serving as Reviewer for technical and as Organizer/Co-organizer of numerous international and national conferences and workshops. In addition, he served as Programme Committee Member for international conferences and workshops.

He was responsible for the formation cycle "Maîtrise de Génie Informatique" in the Faculty of Sciences and Technologies in Tangier since 2006. He is responsible too and Coordinator of Tow Master "DCESS - Systèmes Informatique pour Management des Entreprise" and "DCESS - Systèmes Informatique pour Management des Enterprise". He is Coordinator of the computer science modules and responsible for the graduation projects and external relations of the Engineers Cycle "Statistique et Informatique Décisionnelle" in Mathematics Department of the Faculty of Sciences and Technologies in Tangier since 2007. He participates also in the Telecommunications Systems DESA/Masters, "Bio-Informatique" Masters, and "Qualité des logiciels" Masters in the Faculty of Science in Tetuan since 2002.

He is also Founder and Current Chair of the blinds and partially signet people association. His activity interest focuses mainly on the software to help the blinds and partially signet people to use the ICT, specifically in Arabic countries. He is Founder of the private centre of training and education in advanced technologies AC-ETAT, in Tangier since 2000.

A Novel Route Discovery Mechanism Based on Neighborhood Broadcasting Methods in VANET

Khalid Kandali[1(✉)], Hamid Bennis[1], and Mohamed Benyassi[2]

[1] TIM Research Team, EST of Meknes, Moulay Ismail University,
Meknes, Morocco
kandalikhalid@hotmail.com
[2] LEM2A Laboratory, EST of Meknes, Moulay Ismail University,
Meknes, Morocco

Abstract. Vehicular Ad-hoc Networks (VANET) are considered a promising application in Intelligent Transport Systems (ITS). It is characterized by a high mobility, a high density of nodes, and a very constrained topology that require a more suitable and more reliable routing protocol. Hence, the need to optimize the current routing protocols used in Mobile Ad-hoc Networks (MANET), in order to increase their performance in VANET. Broadcast communication method remains a very important step to adapt the broadcast information to other nodes and to select the routes in better conditions. In this article, we propose a new broadcast approach based on the neighborhood broadcasting methods. Based on this solution, we present an improving Ad-hoc On-Demand Distance Vector Routing (AODV) to reduce the number of dropped packets, and minimize network overhead. The simulation shows encouraging results of our proposed approach.

Keywords: VANET · ITS · AODV · MANET · Broadcast · Routing protocols

1 Introduction

VANET are considered as a subclass of MANET, whose main objective is to improve the Intelligent Transportation System (ITS) [1]. VANET must be adapted to a high mobility of the nodes, a frequent change of the topology, a random disconnection of the nodes, and finally face the blocking of signals transmission because of the buildings in urban environment [2]. All these constraints have motivated researchers to improve VANET to this new environment. The works are focused on routing as it is considered the most important aspect in the field of VANET.

Routing takes care of finding stable routes to forward packets from the source node to the destination. These routes must take the least amount of time and must be more reliable, and less expensive [3]. If a source node is not able to send a message directly to its destination node that does not exist in its communication range, the source node uses intermediate nodes to forward the message to the destination. It broadcasts routing packets to collect information needed to discover routes. This process is very important in routing protocols.

M. Ezziyyani (Ed.): AI2SD 2019, LNNS 92, pp. 1–12, 2020.
https://doi.org/10.1007/978-3-030-33103-0_1

We note well, that the routing in the VANET, is based on three phases:

- Route Discovery Phase: When the source node wants to transmit data to a destination node, it looks in its routing table for a route to this node, otherwise it starts the route discovery phase by sending Route REQuest (RREQ) to all these neighbors. Each node that receives the RREQ packet, forwards it to all its neighbors, if there is not a route to the destination node. Otherwise it generates a Route REPly (RREP) and forwards it back to the source node. The letter receives RREP packets, and selects the shortest route.
- Data Transmission Phase: This phase starts when the source node knows a route to the destination node.
- Route Maintenance Phase: It starts when a link to a node destination is disconnected or down. In this case the node sends a Route ERRor (RERR) to the source node, so that the link to the next node is disconnected.

In Route Discovery Phase, there are two steps. Step of selection the reliable routes, and step of routing packets broadcasting. In the second step, there are several methods that are based on probabilistic, statistical and geometric models [4]. These methods are classified into four sections [5]:

- The simple flooding methods [6]: in which the nodes broadcast and rebroadcast all packets and to all neighboring nodes.
- The probability-based methods [7]: each node calculates a probability based on the topology of the network, to broadcast or drop the packet.
- The area-based methods [7]: or the broadcast and rebroadcast are based on a common transmission distance between the nodes. The rebroadcast here is done if there is a sufficient communication range.
- The Neighborhood-based methods [8]: the broadcast is done according to parameters of mobility of the neighboring nodes.

In this article, we will introduce a new method of routing packets broadcast based on the neighborhood methods. This approach takes into account the position, speed, and direction of other neighboring nodes. This approach is an improvement of another approach proposed in the [9] that will be called IMPR1-AODV. We applied this new method in the AODV protocol to obtain a new routing protocol called IMPR2-AODV. A lot of research works on routing in VANET consider that AODV is the most suitable protocol in this type of network [3].

IMPR2-AODV will be evaluate in terms of packet delivery ratio (PDR) and normalized routing load (NRL), with the original protocol and the IMPR1-AODV protocol [9].

Furthermore, this paper has been divided in five main sections. In addition to Sects. 1 and 2 discusses the work already done in relation with our subject. Section 3 explains our proposed approach IMPR2-AODV. Section 4 shows the performance evaluation. In Sect. 5, we conclude this study and outline some recommendations for future research work.

2 Related Works

We have made a global view in the literature about the proposed solutions that modify the routing algorithms in MANET, in order to optimize these algorithms in VANET environments. The authors based in their work specially on deterministic algorithms like; Simple Flooding, Probability Based Methods, the area-based Methods and the neighborhood-based. Most of the performance comparison works in the literature have shown the weakness of the Probability based methods regarding the number of packets dropped, although this method requires less information, less memory and less computing power.

In the paper [9], the authors proposed two steps of optimizations in route discovery and route selection process in order to improve the stability of the route and to decrease overhead. In the route discovery phase, the nodes whose links are more stable are selected to receive the RREQ packets. The stability of the nodes is measured according to the speed and direction information. In the phase of route selection process, the authors proposed an optimization of the selection of the route when the source node receives several routes to the destination node.

In [10], the authors introduced a new idea called "Reliability Factor" to determine the reliable links between the source node and the destination node in MANET networks. This proposed process considers the reliability factor as the main metric when selecting the route. The reliability factor is calculated according to route expiration time RET and Hop count HC. The route discovery process in the proposed Reliability Factor Based Routing Protocol (RFBRP) aims to minimize routing errors, reducing the number of route discovery requests as well as the computational overhead of each node.

The paper [11], is based on the main operating techniques of AODV to propose the new I-AODV approach. The author has proposed two strategies that work productively to improve network performance. The first strategy improves the control of HELLO messages by sharing information from the neighboring node with other nodes. This helps the protocol in reducing the congestion of network. And the other technique creates an alternative route option for faster delivery of data packets. By combining the information table within the HELLO Message, the node can select the effective node directly for data transmission.

In the paper [12], the authors presented Proactive AODV (Pro-AODV) protocol that uses information from the AODV routing table to minimize congestion in VANET. This improvement is based on Probability Based Methods for broadcasting packets. In Pro-AODV, each node decides to drop or broadcast the packet according to a probability p. This probability depends on the number of entries in its routing table.

To improve the performance of existing on-demand routing protocols, the authors in the paper [13] proposed the Smart Probabilistic Broadcasting (SPB) as a new probabilistic method. This approach is essentially aimed at minimizing the number of RREQ packets during the route discovery phase. The combination of this approach with the AODV protocol has minimized the end-to-end delay and control overhead.

3   Our Proposed Approach

In this paper, we propose a new method for the Route discovery Phase in a routing protocol. This process is about finding the shortest, most reliable, and most stable route. This step starts when a source node wants to transmit data to a destination node, especially when the source node cannot find a route to the destination in its routing table. The main goal behind this method is to reduce the number of dropped packets, and minimize the control overhead.

In this step, the source node usually broadcasts RREQ packets to all its neighbors in order to reach the destination node. This method increases the network overhead and the number of dropped packets, because it supports even the neighbors that will be useless to find the best route.

The uselessness of a neighboring node is in its direction relative to the node sending the packet RREQ. A vehicle traveling in the same direction of another vehicle may remain in its communication range for a long time. The same direction means that the link between the two vehicles becomes more stable. On the other hand, a vehicle traveling in a different direction than another vehicle may quickly leave the communication range of the other vehicle, as it is shown in Fig. 1. Also, a big speed difference between two vehicles can make the neighbor useless, and the link between them less reliable. On the other hand, if two vehicles traveling at a similar speed, remain in the communication range for a long time. So, the direction and the speed play a very important role in the construction of the stability of the routes.

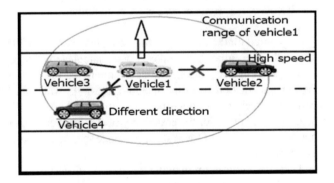

Fig. 1. Neighbor selection in our approach.

Density, mobility, and traffic load are considered important parameters that affect the performance of a VANET network. Vehicles move randomly on the roads, which implies a frequent change in the topology of the network. This change can increase the density of the nodes in a certain moment, as it can decrease this density in another moment. When a source vehicle wants to transmit data to a destination vehicle, it begins the route discovery procedure, by sending RREQ packets to all its neighboring vehicles. While the source vehicle normally must select the neighbors, who can serve it to find the better routing path, and neglect the vehicles that will be useless in this process.

To achieve these objectives, we proposed a new algorithm for broadcasting RREQ packets to neighboring nodes, using mobility parameters from each node of the network. This algorithm selects the neighboring nodes involved in the routing. It supports threshold number of neighbors at startup.

Selection of Nodes with Stable Weight-Based in Route Discovery Process
The first step in this approach is to calculate the weight link of each neighbor.

To calculate the weight between two nodes, the following formula was used in [9]:

$$LW = W_V \times |V_1 - V_2| + W_\theta \times |\theta_1 - \theta_2| \tag{1}$$

Or, W_V represents the speed weight factor, and W_θ represents the direction weight factor.

V_1 represents the speed of the node N_1, and θ_1 is its direction. Whereas V_2 represents the speed of node N_2 and θ_2 is its direction. Speed and direction information can be obtained from the GPS information of each node.

We have modified this formula, by adding the distance between two nodes N_1 and N_2.

We will have the following formula 2:

$$LW = W_D \times D + W_V \times |V_1 - V_2| + W_\theta \times |\theta_1 - \theta_2| \tag{2}$$

Or, D is the distance between the two nodes, and W_D represents the distance weight factor.

As mentioned in [14], we can calculate the direction angle θ of each node. Let two nodes N_1 and N_2 having a position (x_1, y_1), (x_2, y_2) at time t and at time t + Δt are (x'_1, y'_1), (x'_2, y'_2), respectively, as shown in Fig. 2:

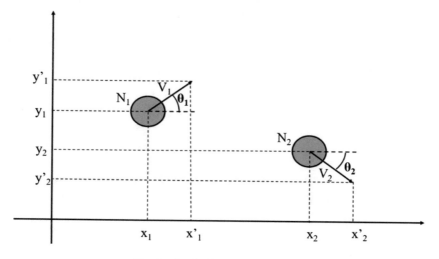

Fig. 2. Angle direction calculation.

The direction angle θ of each node is given by the following formula:

$$\theta = arctan\left(\frac{\Delta y}{\Delta x}\right) \tag{3}$$

With:

$$\begin{cases} \Delta x = x' - x \\ \Delta y = y' - y \end{cases} \tag{4}$$

And the distance between 2 nodes is calculated according to the following formula:

$$D = \sqrt{(x_2 - x_1)^2 + (y_2 - y_1)^2} \tag{5}$$

So, the source vehicle must select the neighbors to which it will transmit the RREQ packets. It calculates in the first time the number of neighboring nodes. If the number of neighboring vehicles for the source vehicle is less than the threshold number of the neighbors, then the source node will broadcast the packet RREQ to all these neighbors. If the number of neighbors is greater than the threshold number, the source node will choose the threshold number of neighbors among all its neighbors. To make this choice, the source node will calculate the link weight of each neighboring node, using its information about speed, direction and position, and it will just consider the neighbors who have the least weight of the link to make transmit the RREQ packet. When the neighbor node receives the packet RREQ, it tests, if it is the destination node or it has a route to the destination node, it generates a packet RREP. Otherwise, it broadcasts the packet RREQ to its neighboring nodes. The intermediate node updates its routing table, if it receives a better route.

The totality of this algorithm is shown in Fig. 3:

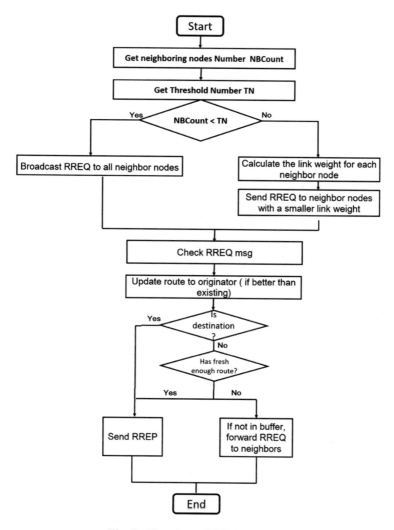

Fig. 3. Flowchart of IMPR2-AODV

This algorithm will be repeated for all other intermediate nodes that received the RREQ packet. Consequently, vehicles that have a different direction, or a very high speed compared to the source vehicle, will not necessarily be involved in sending the RREQ packets.

Algorithm of RREQ Flooding:

```
Get threshold number TN;
Let n be the number of neighbors for the node X that
broadcasts the packet RREQ;
If (n <= TN) then
      Node X will broadcast the RREQ packet to all its n
      neighbor;
Else {
      For i from 0 to n
          {
          Calculate the link weight value between the X
          node and the neighbor node I;
          Let Ω be the set of neighboring nodes with
          their link weights;
          }
      End for
      While (j <= TN) then
          {
          Send RREQ to the node I_min(Ω);
          Ω=Ω/I_min(Ω);
          increment j by 1;
          }
      End while
      }
End if
End Algorithm
```

4 Performance Evaluation

The effectiveness of our approach has been evaluated on the NS 2 simulator. This evaluation consists of comparing our proposed approach IMPR2-AODV with IMPR1-AODV and also with the original AODV protocol. Our approach is to improve Packet Delivery Ratio, and decrease Overhead Control. The simulation is done according to several scenarios, varying the number of nodes and the speed of the nodes in the network. The simulation takes place over an area of 1000 m × 1000 m. Nodes choose random starting points and move in different directions, and stop for a period of time. The roads are composed of 2 lanes.

The first scenario consists of changing each time the number of nodes {25, 50, 75, 100, 125, 150}, while in the second scenario, we set the number of nodes to 50, and to increase the speed of the nodes in a regular way {10, 20, 30, 40, 50, 60 m/s}.

We put connections between nodes (sources and destinations), which move in the same direction and in different directions.

According to several tests, better results have been obtained when the value of threshold is initially set to 10. Also, according to several simulations, the most optimal values for the different weight factors are: $W_D = 0.2$; $W_v = 0.3$ and $W_\theta = 0.5$. The simulation was run for 10 times in a different way and the average value of the results obtained was taken.

The following table represents all the parameters used in this simulation:

Table 1. Parameters of simulation.

Parameter	Value
Simulator	NS2
Simulation time	900 s
Area size	1000 m × 1000 m
Number of nodes	25, 50, 75, 100, 125, 150
Mobility model	Random way point
Speed distributions	10, 20, 30, 40, 50, 60 m/s
Transmission range	250 m
Propagation model	Two-ray ground
Traffic type	Constant bit rate (CBR)
Packet rate	4 pkts/s
Packet size	512 bytes

4.1 Effect of Number of Nodes

Figure 4 represents the impact of changing the number of nodes on the Packet delivery ratio (PDR). The PDR is calculated by dividing the number of packets delivered to a destination node by the number of packets sent by the source node [15]. This value is expressed in %. Our approach has responded favorably by keeping PDR value higher than other approaches. AODV and IMPR1-AODV lose the value of PDR more than IMPR2-AODV when the number of nodes increases.

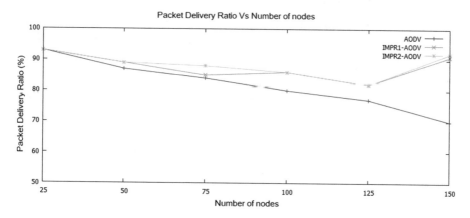

Fig. 4. Packet delivery ratio vs number of nodes.

Normalized Routing Load (NRL) shows the number of routing packets needed to transmit data packets from a source node to a destination node. its value is calculated by dividing the number of routing packets by the number of data packet [16]. It can be seen in the Fig. 5 that the NRL reached by the three protocols increases as the number of nodes increases. The figure also shows that the performance difference between the three routing protocols becomes more apparent as the number of nodes increases. Indeed, when the numbers of nodes are greater than 125, IMPR2-AODV keeps a value less than the other protocols. Most of the generated data packets are deleted as a result of the direction, position, and speed of the nodes.

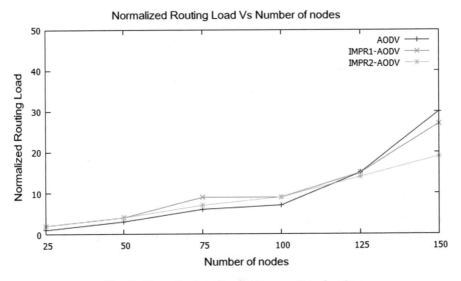

Fig. 5. Normalized routing load vs number of nodes

4.2 Effect of Speed of Nodes

Figure 6 illustrates the impact of the speed of nodes on the Packet Delivery Ratio. Here, PDR is high compared to IMPR1-AODV and traditional AODV. As shown in Fig. 6, our proposed solution is remarkably good in a network with a high speed.

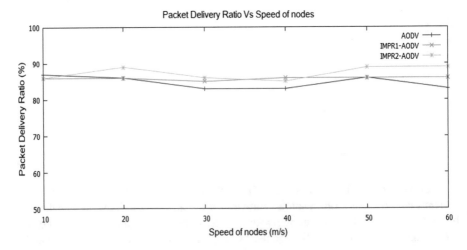

Fig. 6. Packet delivery ratio vs speed of nodes

While Fig. 7 shows the impact of increasing the speed of nodes on Normalized Routing Load. The lower the value of NRL, the better is the performance of the protocol. Our approach shows good results when the speed of the nodes exceeds 40 m/s.

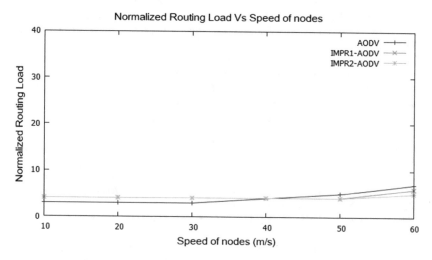

Fig. 7. Normalized routing load vs speed of nodes

5 Conclusion and Perspectives

In this article, we have proposed a new broadcasting method approach for routing protocol in VANET. This solution based on Neighborhood-Broadcasting takes into consideration the distance, speed, and direction of neighboring nodes, to broadcast routing packets. Only neighbors who have the most stable links will receive and rebroadcast the packets. We tested our approach by comparing it with another approach proposed in the literature and with the original AODV protocol. We have achieved satisfactory results. Our approach is favorable compared to other approaches in terms of PDR and NRL.

The broadcasting method remains the first step in the route discovery. We aim in our future work to improve the route selection approach as the second step. We are looking forward to proposing a new algorithm in order to minimize the average end to end delay.

References

1. Ennaciri, A., Khadim, R., Erritali, M., Mabrouki, M., Bengourram, J.: Contribution to the improvement of the quality of service over vehicular ad-hoc network. In: Advanced Intelligent Systems for Sustainable Development (AI2SD 2018), vol. 915, pp. 170–190. Springer International Publishing, Cham (2019)

2. Kumar, Y., Kumar, P., Kadian, A.: A survey on routing mechanism and techniques in vehicle to vehicle communication (VANET). Int. J. Comput. Sci. Eng. Surv. **2**(1), 135–143 (2011)
3. Kandali, K., Bennis, H.: Performance assessment of AODV, DSR and DSDV in an urban VANET scenario. In: Advanced Intelligent Systems for Sustainable Development (AI2SD 2018), vol. 915, pp. 98–109. Springer International Publishing, Cham (2019)
4. Mkwawa, I.-H.M., Kouvatsos, D.D.: Broadcasting methods in MANETS: an overview. In: Kouvatsos, D.D. (ed.) Network Performance Engineering, vol. 5233, pp. 767–783. Springer, Heidelberg (2011)
5. I.S. Committee: Wireless LAN Medium Access Control (MAC) and Physical Layer specifications. IEEE 802.11 Standard. IEEE, New York (1997)
6. Ho, C., Obraczka, K., Tsudik, G., Viswanath, K.: Flooding for reliable multicast in multi-hop ad hoc networks. In: International Workshop in Discrete Algorithms and Methods for Mobile Computing and Communication, pp. 64–71 (1999)
7. Tseng, Y., Ni, S., Chen, Y., Sheu, J.: The broadcast storm problem in a mobile ad hoc network. In: International Workshop on Mobile Computing and Networks, pp. 151–162 (1999)
8. Lim, H., Kim, C.: Multicast tree construction and flooding in wireless ad hoc networks. In: Proceedings of the ACM International Workshop on Modeling, Analysis and Simulation of Wireless and Mobile Systems, MSWIM (2000)
9. Ding, B., Chen, Z., Wang, Y., Yu, H.: An improved AODV routing protocol for VANETs, pp. 1–5 (2011)
10. Khan, S.M., Nilavalan, R., Sallama, A.F.: A novel approach for reliable route discovery in mobile ad-hoc network. Wireless Pers. Commun. **83**(2), 1519–1529 (2015)
11. Mittal, S., Bisht, S., Purohit, K.C., Joshi, A.: Improvising-AODV routing protocol by modifying route discovery mechanism in VANET. In: 2017 3rd International Conference on Advances in Computing, Communication & Automation (ICACCA) (Fall), Dehradun, pp. 1–5 (2017)
12. Kabir, T., Nurain, N., Kabir, M.H.: Pro-AODV (proactive AODV): simple modifications to AODV for proactively minimizing congestion in VANETs. In: 2015 International Conference on Networking Systems and Security (NSysS), Dhaka, Bangladesh, pp. 1–6 (2015)
13. Bani Yassein, M., Bani Khalaf, M., Al-Dubai, A.Y.: A new probabilistic broadcasting scheme for mobile ad hoc on-demand distance vector (AODV) routed networks. J. Supercomput. **53**(1), 196–211 (2010)
14. Hadded, M., Muhlethaler, P., Laouiti, A., Azzouz Saidane, L.: A novel angle-based clustering algorithm for vehicular ad hoc networks. In: Laouiti, A., Qayyum, A., Mohamad Saad, M.N. (eds.) Vehicular Ad-Hoc Networks for Smart Cities, vol. 548, pp. 27–38. Springer, Singapore (2017)
15. Nabou, A., Laanaoui, M.D., Ouzzif, M.: The effect of transmit power on MANET routing protocols using AODV, DSDV, DSR and OLSR in NS3. In: Advanced Intelligent Systems for Sustainable Development (AI2SD 2018), vol. 915, pp. 274–286. Springer International Publishing, Cham (2019)
16. Laqtib, S., El Yassini, K., Hasnaoui, M.L.: Performance evaluation of multicast routing protocols in MANET. In: Advanced Intelligent Systems for Sustainable Development (AI2SD 2018), vol. 915, pp. 847–856. Springer International Publishing, Cham (2019)

Access Control Models for Smart Environments

Yasmina Andaloussi[1]([⊠]), Moulay Driss El Ouadghiri[1],
and Ziad Saif Mohammed Al Robieh[2]

[1] My Ismail University, Meknes, Morocco
andyasmina.dmelouad@gmail.com
[2] Taiz University, Taiz, Yemen
ziadrh@yahoo.com

Abstract. The Internet of Things (IoT) is the extension of the internet to our daily life where common objects, from toothbrush to watches, collect information and interact with their environments with no or little human intervention. These different smart objects interact directly with their physical environments. They collect and transfer sensitive and private data from various users. This puts security and privacy issues at the forefront: the ability to manage the digital identity of millions of people and billions of devices is fundamental for success. As most of the information contained in IoT environment may be personal or sensitive data, there is a requirement to support anonymity and restrain access to information. This paper will focus on access control and authentication mechanisms as well as supporting the cryptography algorithms in constrained devices.

Keywords: IoT · Access control · Security · Privacy · Capability-token · Distributed-capability based access control

1 Introduction

Cryptography is of course a first order technique for achieving this goal, but it's applicability to IoT is challenging: smart objects have limited computing capabilities, limited storage, and often run on batteries. These issues are major constraints for the implementation of current cryptographic mechanisms in IoT environments. Current research is focused on finding lightweight cryptographic solutions [4]. These solutions consist in adapting conventional cryptographic algorithms for IoT objects and devices or in finding new ones that are equally effective in terms of security, execution time and energy consumption. In this paper, we will survey and compare the different algorithms that are used to enforce access-control.

Authentication and access control [1–3] are key security mechanisms for the large adoption of IoT. These are necessary to gain the trust of users and to develop wide scale secure system. There are several models for access control related to intelligent environments, the most current are role-based access control (RBAC), attribute-based access control (ABAC) and capability-based access control (CapBAC).

M. Ezziyyani (Ed.): AI2SD 2019, LNNS 92, pp. 13–18, 2020.
https://doi.org/10.1007/978-3-030-33103-0_2

2 Background

RBAC and ABAC have been developed first. In RBAC [5], Users are assigned to roles, and security policies grant rights these roles. Traditional access control models like RBAC do not take into consideration contextual information (time, location …). In order to provide a more flexible access mechanism, the ABAC (Attribute-Based Access Control) model [5] has been proposed, in which authorization is based on attributes that the user has to prove and take into account context information. The privileges are granted to users by combining all attributes together. Among the principal advantages of ABAC that the requestors should not be known a priori by targets, which is a great flexibility compared to RBAC. But, the entities must agree on the attributes, which is difficult to accomplish in large and distributed systems. In the context of a fully distributed approach, as is often the case for IoT, the feasibility of applying traditional access control models such as RBAC or ABAC is complex and disputed.

For these reasons, this paper will mainly cover the CapBac model that was proposed as a feasible approach for IoT scenarios [3] and is supported by constrained devices. This approach is based on the concept of capability which is defined in [4] as "token, ticket, or key that gives the possessor permission to access an entity or object in a computer system". This token contains the privileges that will be granted to the entity that holds the token. This token must be tamper-proof and identified unequivocally to be considered in a real environment. Moreover, it is necessary to have a cryptographic mechanism which is supported even by the constrained devices.

3 Discussion About Access Control Models

Nowadays, there is a myriad of access control models that are applied to different Internet of Things scenarios in which security is required.

In the following, we give a brief description of the most popular models, which are deployed in such scenarios.

In the Mandatory Access Control (MAC) model [1] the administrator of the system give permissions for subject to access object. The model assigns security labels to subjects and objects, and it is independent of the user operations, only the administrator can modify object security labels. MAC models are difficult and expensive to implement and maintain, its usage is usually limited to military applications, and this is why MAC models are not used as access control system.

In the Discretionary Access Control (DAC) models [2], the access to resources is maintained by users, which can grant permissions to their resources by being included in Access Control Lists (ACL).

Each entry in the access control list gives users (or group of subjects) permissions to access resources.

The permissions are usually stored by objects. Unlike in MAC, where permissions are given in predefined policies by the administrator, in DAC, permissions are given by users which decide the access rights to the resources they belong. DAC is adopted by current operation systems based on UNIX, FreeBSD, and Windows.

Moreover, in the Role-Based Access Control (RBAC) model [3], users are assigned to roles, and the security policies grant rights to roles rather than to users. Since the users are associated to roles.

RBAC allows creating hierarchies of permissions and inheritance. Nonetheless, RBAC has some problems since the administrative issues of large systems where memberships make administration potentially cumbersome.

Traditional access control models like MAC, DAC and RBAC do not take into account additional parameters such as resource information and dynamic information (such as time, location ...). In order to provide a more flexible mechanism, the Attribute-Based Access Control (ABAC) model [4] was proposed, in which authorization decisions are based on attributes that the user has to prove (e.g.: age, location, roles, etc.). One of the main advantages of ABAC is requesters do not have to be known a priori by targets, providing a higher level of flexibility for open environments, compared to RBAC models.

Nevertheless, in ABAC everyone must agree on a set of attributes and their meaning when using ABAC, which is not easy to accomplish.

The AuthoriZation-Based Access Control (ZBAC) [5–7] model uses authorization credentials, which are presented along a request to make an access control decision. Unlike ABAC and RBAC systems, in which the user submits an authentication along with the service request, in ZBAC systems, the user submits an authorization along with the request.

IoT scenarios imposes significant restrictions on privacy and access control, tradition access control approaches solutions were not designed with these aspects.

4 Issues and Problems

In this section we present the approaches related to access control, highlighting their main advantages and drawbacks.

Centralized Approach
In a centralized approach, all access control logic is externalized into a central entity responsible for filtering access requests based on their authorization policies. The end-devices (sensors ...) play a limited role to as information providers (Fig. 1).

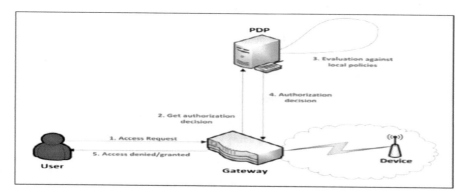

Fig. 1. Centralized approach [3]

This centralized approach does not take into consideration constraints of resources, because the access control logic is located in an entity without constraints of resources. However, the centralized approach has major problems. First, the end-device is not taken into consideration in access control decisions. Second, the access control logic is located in one entity, so it becomes a single point of failure; consequently, any vulnerability might compromise all the system.

Hybrid Approach

The end-devices are not passive entities; they participate in the access control decisions such as presented in the Fig. 2. The limit of this approach is the ability of the end-devices to taking into account contextual information (location, time …) for making decisions.

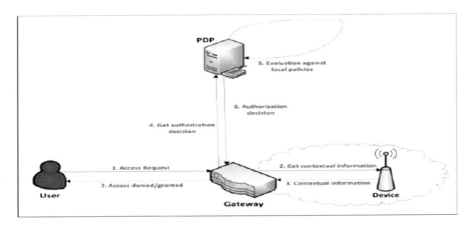

Fig. 2. Hybrid approach [3]

The limit of this approach is the delays that can be introduced in transmission the contextual information from the end-device to central entity. So the information loses the importance in the process of authorization decision.

Distributed Approach

In the distributed architectures, the end-device is a smart thing that is enable to obtain, process and send information to other services and devices. The devices are able to take authorizations decisions without the need of central entities (Fig. 3).

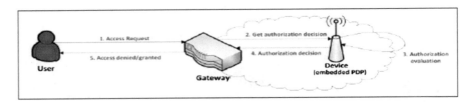

Fig. 3. Distributed approach [3]

This approach presents interesting features and it is more suitable for the IoT.

Distributed Capability-Based Access Control

DCapBAC has been postulated as a feasible approach to be deployed on IoT scenarios even in the presence on devices with resource constraints. The key concept of this approach is the concept of capability, which was originally introduced by [5–7] as "token, ticket, or key that gives the possessor permission to access an entity or object in a computer system". This token is usually composed by a set of privileges which are granted to the entity holding the token. Therefore, it is necessary to consider suitable cryptographic mechanisms to be used even on resource-constrained devices which enable an end-to-end secure access control mechanism.

In a DCapBAC scenario, an entity wants to access a resource of another entity. Usually, an issuer generates a token for the subject specifying which privileges it has. Thus, when the entity attempts to access a resource hosted in the other entity, it attaches the token which was generated by the issuer. Then, the entity evaluates the token granting or denying access to the resource (Fig. 4).

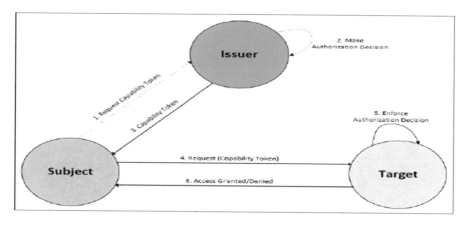

Fig. 4. Scenario of distributed capability based access control [5]

5 Conclusion

Security is a challenge for IoT. IoT requires limit protection and privacy enhancement through the optimized use of cryptographic primitives such as ECC and authentication protocols to differentiate between permitted and unauthorized entities.

Safety for IoT is a major requirement as the inclusion of intelligent systems is extended to real-life scenarios such as hospitals, home automation and smart cities. Therefore, the theft of private and intimate information can cause significant damage.

References

1. Ye, N., Zhu, Y., Wang, R., Malekian, R., Qiao-Min, L.: An efficient authentication and access control scheme for perception layer of internet of things. J. Appl. Math. Inf. Sci. **8**, 1617 (2014)
2. Khalid, X.H., Le, M., Sankar, R., et al.: An efficient mutual authentication and access control scheme for wireless sensor networks in healthcare. J. Netw. **6**, 355–364 (2011)
3. Liu, J., Xiao, Y., Chen, C.L.P.: Authentication and access control in the internet of things. In: Proceedings 32nd International Distributed Computing Systems Workshops, Macau, pp. 588–592 (2012)
4. Eisenbarth, T., Kumar, S., Paar, C., Poschmann, A., Uhsadel, L.: A survey of lightweight-cryptography implementations. IEEE Des. Test Comput. **24**(6), 522–533 (2007)
5. SOCIOTAL: Creating a socially aware citizen-centric Internet of Things, Specific Targeted Research Projects (STReP) (2014)
6. Andaloussi, Y., et al.: Access control in IoT environments: feasible scenario. Procedia Comput. Sci. **130**, 1031–1036 (2018)
7. Andaloussi, Y., et al.: Model-based security implementation on internet of things. In: (AI2SD 2018) Advanced Intelligent Systems for Computing Sciences (2018)

Ambulance Detection System

Baghdadi Sara[1]([⊠]), Aboutabit Noureddine[1], and Baghdadi Hajar[2]

[1] IPIM Laboratory, ENSA, Khouribga, Morocco
sara92.baghdadi@gmail.com
[2] STI Laboratory, ENSA, Khouribga, Morocco

Abstract. In this paper, we aim to develop a computer vision system to detect robustly the ambulances observed from a static camera. Robust and reliable ambulance detection plays an important role for priority systems. These aim to find the shortest possible paths of ambulances till their destination by managing signaling networks with traffic lights. If the emergency vehicle gets stuck in a traffic jam and its arrival at the incident location is delayed, it can cause the loss of lives and property. The approach is based on two main stages: Feature extraction step and classification step. For the first step, we used many descriptors to extract features, such as HOG (Histogram of Oriented Gradient), LBP (Local Binary Patterns), and Gabor filter. Then, we classified these features using machine learning algorithms like SVM (Support Vector Machines) and kNN (K-nearest neighbor). To evaluate the performance of our constructed models, we calculated many metrics: True Positive, False Positive, False Negative, True Negative, accuracy and runtime. In fact, our achieved results show that the ambulance detection system can be successfully exploited for the overall system.

Keywords: Machine learning · Ambulance detection · Emergency vehicles · Medicine · Classification · Feature extraction

1 Introduction

Nowadays, with the increase of vehicles on road, ambulances must wait in traffic which delays their arrival at their destination [1]. The excellence of the emergency service depends on how fast the emergency vehicles can reach the incident location. This situation is often happening because of increase of vehicular population [2] and sometimes lack of cooperation from civilians [3]. If the emergency vehicle gets stuck in a traffic jam and its arrival at the incident location is delayed, it can cause the loss of lives and property [4].

There is a need of a system that prioritize ambulances by managing traffic signal networks with traffic lights. Automatic ambulance detection system plays an important role for priority systems. It can be also used in monitoring traffic rules and regulation systems to allow, for example, accessing speed limits for these emergency vehicles without sending an alert for violation law. Ambulance detection system can be also used in self-guided vehicles to alert the driver that an emergency vehicle is approaching.

© Springer Nature Switzerland AG 2020
M. Ezziyyani (Ed.): AI2SD 2019, LNNS 92, pp. 19–25, 2020.
https://doi.org/10.1007/978-3-030-33103-0_3

In this work, we will present a concise overview on machine learning models that we used to build this system. These models are based on two main stages: feature extraction step and classification step (Fig. 1). Several techniques can be used for each stage. In the first step, we used three different kinds of descriptors: LBP, HOG and Gabor filter. In the second step, we used two classifiers: SVM and kNN.

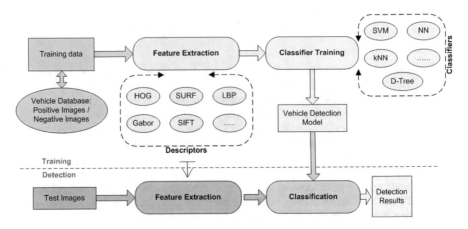

Fig. 1. Vehicle detection system [5]

The next section presents the related research to ambulance detection system.

2   Related Work

In the literature, researchers have suggested various solutions to prioritize ambulances but without using Machine Learning solutions. The authors in [3] designed a warning system, EVRS, for saving the life of people. In fact, this system could alert the driver of a private vehicle that an emergency vehicle is approaching. The EVRS guides the driver in a cleared path by avoiding delay in the traffic. The authors in [6] also presented a solution for traffic management of emergency vehicles that adjusts the traffic signals dynamically and also recommend drivers required behavior changes, driving policy changes and exercise necessary security controls. In [7], the authors proposed a solution to control traffic for clearing ambulances, detect stolen vehicles and control traffic jams. To do this, radio frequency identification (RFID) tags are affixed to vehicles to assist in counting the number of vehicles on a particular path, detecting the stolen vehicle and delivering a message to the police control room. In addition, it communicates with the traffic controller to prioritize ambulances using ZigBee modules. In [8], a system has been designed to detect the incoming direction of an emergency vehicle. Acoustic detection methods based on a cross microphone array have been implemented. Source detection based on an estimation of time delay has been shown to outperform sound intensity techniques, although both techniques perform well for the application. The authors in [2] use the fusion of Internet of Things (IoT) and Vehicular Ad Hoc Network (VANET) to investigate an Intelligent Traffic

Management System (ITMS). Their solution could navigate ambulances to find the shortest possible paths and present a counter measure to get rid the problem of the traffic light system when it is hacked during its operation.

In [1], the authors design an advanced adaptive traffic control system that enables faster emergency services response in smart cities while maintaining a minimal increase in congestion level around the route of the emergency vehicle. This can be achieved with a Traffic Management System (TMS) capable of implementing changes to the road network's control and driving policies. In [9], the authors propose a system for reducing traffic congestion using image processing by detecting blobs and tracking them. The system detect vehicles through images and also help emergency vehicles stuck in traffic to clear the route by using Bluetooth. [4] presents an approach for scheduling emergency vehicles in traffic. The approach combines measuring the distance between the emergency vehicle and an intersection using visual sensing methods, vehicle counting and time sensitive alert transmission within the sensor network. The distance between the emergency vehicle and the intersection is calculated for comparison using Euclidean distance, Manhattan distance and Canberra distance techniques.

3 Ambulance Detection Methods

This section presents all methods used in our work. We start by defining the descriptors and then the classifiers.

3.1 Descriptors

Histogram of Oriented Gradient HOG. Introduced in 2005 by Dalal and Trigg, the HOG is a feature descriptor used to detect objects [10]. As seen in Fig. 2, the image is segmented into small regions called cells, which are interconnected. The HOG directions are compiled for each pixel of these cells using the gradient extraction operator. A histogram is formed in different orientations according to the accumulated value of the gradient, the HOG characteristics of each cell are extracted to the series and a one-dimensional feature vector of the vehicle image is obtained.

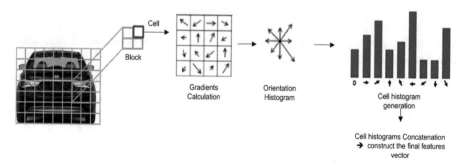

Fig. 2. HOG features [5]

Local Binary Model (LBP). The LBP operator is a texture operator which labels the pixels of an image by thresholding the neighborhood of each pixel and considers, as a binary number, the result. The LBP feature vector is created as shown in Fig. 3.

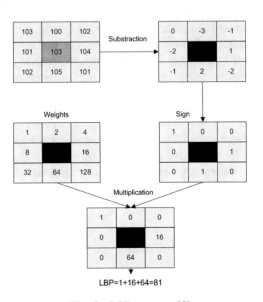

LBP=1+16+64=81

Fig. 3. LBP operator [5]

By applying LBP operator to one image, one pattern map can be computed. The pattern map is divided into many blocks and the histogram computed in each block is concatenated together to form the description of the input image [11].

3.2 Classifiers

Support Vector Machines (SVM). Was initially proposed by Cortes and Vapnik [12] in 1995. The principle of SVM classification is to find a hyper plane, which makes the classification of multidimensional space maximized. In the linear binary classification, the category labels are only two values, respectively, 1 and −1. In order to minimize errors, it is necessary to maximize the geometrical margin. SVM uses multiple nuclei such as the polynomial kernel, the linear kernel, and the Gaussian Radial Basic Function (RBF) core [12, 13].

K-Nearest Neighbor (kNN). Was proposed by Cover and Hart. This is the extension of the minimum distance method and the nearest neighbor method. It calculates the distance between the test sample x and all the learning samples and sets the distances in decreasing order. Then, it selects k training samples that are closest to the test sample x and counts the category of k selected learning samples. The test sample belongs to the category with the highest number of votes in the same category [14].

4   Experiments

For experiments, we implemented these algorithms in Matlab on a Lenovo ThinkPad with a processor Intel Core i5, 7th Generation, CPU @2.50 GHz 271 GHz, RAM 8Go. We've gathered a data set of ambulance images from ImageNet database [15] having 900 images. This data set includes two classes of images, the first class contains the positive images (ambulance images from different views Fig. 4), and the other class contains the background. This step is very important because the quality and quantity of data that we gather will directly determine how good the predictive model can be. Furthermore, the parameters are set as follows: the kernel function of SVM is the linear kernel, the parameter k of kNN classifier is set to 1 (by default) and the distance metric is Euclidean (by default).

Fig. 4. Positive images

To evaluate the performance of our models, we calculated many parameters: True Positive, False Positive, False Negative, True Negative and runtime. Evaluation allows us to test our models against data that has never been used for training. From this, we calculated, True Detection rate, Extra Detection rate, and Missed Detection rate as follows:

True Detection = (TP + TN)/(TP + TN + FP + FN)

Extra Detection = FN

Missed Detection = FP

As shown in the Table 1, LBP Gabor Magnitude and Gabor Phase features need less time to be extracted than HOG features. Histograms need more time to compute. We observed that LBP+SVM is the fastest model.

Table 1. Average runtime and detection rates of each model

Vehicle detection methods		Run-time/image (second)	Training data			Test data		
			True detection	Number of extra detection (average)	Number of missed vehicles (average)	True detection	Number of extra detection (average)	Number of missed vehicles (average)
Combinaison HOG+SVM		0.049230	100	0	0	97.78	13.8	0
Combinaison HOG +KNN	k = 1 & Distance = 'euclidean'	0.187373	100	0	0%	97.22	0	3.1%
Combinaison LBP+SVM		0.015779	99.86	0	0.2	100	0	0
Combinaison LBP+KNN	k = 1 & Distance = 'eucliean'	0.023226	100	0	0	100	0	0
Combinaison Gabor Magnitude+SVM		0.021879	100	0	0	100	0	0
Combinaison Gabor Magnitude+KNN	k = 1 & Distance = 'euclidean'	0.028630	100	0	0	100	0	0
Combinaison Gabor Phase+SVM		0.021372	93.75	25.9	2.5	98.33	0	1.9
Combinbaison Gabor Phase+KNN	k=1 & Distance ='euclidean'	0.029105	100	0	0	97.78	0	2.5

As shown, the descriptors LBP and Gabor Magnitude perform much better than the other descriptors. Texture features like LBP are robust to illumination changes and shadows. All descriptors work very well in combination with SVM classifier.

Basing on the results with test data, the order of the best models according to the accuracy is as follows: LBP+SVM, LBP+KNN, Gabor Magnitude+SVM, and Gabor Magnitude+KNN.

We observed that Gabor Phase+SVM, Gabor Phase+kNN, and HOG+kNN produce false positive detections, caused by illumination changes and shadows. HOG+SVM produce false negative detections due to camouflage (when background and foreground share similar colors) [5].

5   Conclusion

The escalation of vehicle density and population growth increase the response time of emergency vehicles, which is only defined as the period between the received time by an emergency service provider and the arrival time of an ambulance at the emergency site. Reducing this response time by just one minute increases the survival rate for patients with cardiac arrest by 24%. Our main motivation is to develop a system that allows ambulances to bypass the dense traffic and reach their destinations on time. Thus, ambulance detection system is an integral part of this overall system.

In this article, we presented a comparative study of ambulance detection methods. We compared the constructed models using the classifiers: SVM and kNN with the descriptors HOG, LBP and Gabor filter. From the comparison results, we concluded that the descriptors LBP and Gabor Magnitude provide the best results with an accuracy of 100%. LBP with SVM classifier is the fastest one with a runtime per frame of about 0.015779 s. In fact, our achieved results show that the ambulance detection system can be successfully exploited for the previous applications.

For future directions, we will develop the global priority system using artificial intelligence techniques. We will extend our study by introducing various emergency vehicles such as ambulances, police cars, fire engines, etc.

References

1. Djahel, S., et al.: Reducing emergency services response time in smart cities: an advanced adaptive and fuzzy approach. In: 2015 IEEE 1st International Smart Cities Conference ISC2 2015 (2015). https://doi.org/10.1109/ISC2.2015.7366151
2. Ranga, V., Sumi, L.: Intelligent traffic management system for prioritizing emergency vehicles in a smart city. Int. J. Eng. **31**, 2 (2018). https://doi.org/10.5829/ije.2018.31.02b.11
3. Swathi, T., Mallu, B.V.: Emergency vehicle recognition system. Int. J. Eng. Trends Technol. **4**, 987–990 (2013)
4. Nellore, K., Hancke, G.P.: Traffic management for emergency vehicle priority based on visual sensing. Sensors **16**(11), 1–28 (2016). https://doi.org/10.3390/s16111892
5. Baghdadi, S., Aboutabit, N.: Illumination correction in a comparative analysis of feature selection for rear-view vehicle detection. Int. J. Mach. Learn. Comput. (IJMLC)
6. Djahel, S., Salehie, M., Tal, I., Jamshidi, P.: Adaptive traffic management for secure and efficient emergency services in smart cities. In: 2013 IEEE International Conference on Pervasive Computing and Communications Workshops (PERCOM Workshops), pp. 340–343. IEEE (2013)
7. Sundar, R., Hebbar, S., Golla, V.: Implementing intelligent traffic control system for congestion control, ambulance clearance, and stolen vehicle detection. IEEE Sens. J. **15**(2), 1109–1113 (2015)
8. Fazenda, B., et al.: Acoustic based safety emergency vehicle detection for intelligent transport systems. In: ICCAS-SICE, 2009, pp. 4250–4255 (2009)
9. Srinivasan, V., et al.: Smart traffic control with ambulance detection. In: IOP Conference Series Materials Science and Engineering, vol. 402, no. 1 (2018). https://doi.org/10.1088/1757-899X/402/1/012015
10. Wang, H., Zhang, H.: Hybrid method of vehicle detection based on computer vision for intelligent transportation system. Int. J. Multimedia Ubiquit. Eng. (IJMUE) **9**(6), 105–118 (2014)
11. Xie, S., Shan, S., Chen, X., Chen, J.: Fusing local patterns of Gabor magnitude and phase for face recognition. IEEE Trans. Image Process. **19**(5), 1349–1361 (2010)
12. Kim, K-j: Financial time series forecasting using support vector machines. Neurocomputing **55**(1), 307–319 (2003)
13. Schlkopf, B., Tsuda, K., Vert, J.-P. (eds.): Kernel Methods in Computational Biology. MIT Press, Cambridge (2004)
14. Lin, S., Zhao, C., Qi, X.: Comparative analysis of several feature extraction methods in vehicle brand recognition. In: 10th International Conference on Sensing Technology (ICST), Nanjing, China (2016)
15. ImageNet Database: http://www.image-net.org

An AODV-Based Routing Scheme with Layered Clustering to Enhance Energy Efficiency and QoS in WSNs

Alaa Aadri$^{(\boxtimes)}$ and Najlae Idrissi

Faculty of Sciences and Techniques, Information Processing Decision
Support Laboratory, University of Sultan Moulay Slimane,
P.B. 523, Beni Mellal, Morocco
{a.aadri,n.idrissi}@usms.ma

Abstract. The evolution of computer networks and the development of wireless technology have opened up new opportunities in the field of telecommunications and intelligent environments. Wireless sensor networks are large-scale ad hoc networks that consist of several sensors spatially dispersed to measure environmental parameters and to send the collected data to the Sink through a wireless connection. In this context, minimizing energy consumption represents the most important design factor since each sensor node is powered by a limited and generally irreplaceable source of energy. This paper presents a new energy-efficient data routing approach specifically designed for WSNs based on the combination of an extended AODV routing scheme and on network hierarchization in several layers while maintaining low complexity, high scalability and good QoS. Evaluated under the Network Simulator (NS-2), simulation results showed the effectiveness of the proposed approach in terms of energy efficiency and network lifetime.

Keywords: WSNs · Data routing · Network hierarchization · Energy consumption · Load balancing · Network lifetime · QoS

1 Introduction

Advances in the field of radio communications and intelligent environments have facilitated the use of wireless sensor networks considered as a key of the Internet Of Things where nodes consist of a large number of micro-sensors randomly dispersed across the sensing field to collect and transmit environmental data in an autonomous way through a multi-hop routing to the sink node (Fig. 1).

Several routing approaches have been proposed in the literature with the main objective to achieve good data transmission and optimal use of network resources. In this work, we are particularly interested in proposing a routing approach that meets the challenges imposed by the limited energy resource in order to maximize network lifetime and ensure better QoS. The proposed solution combines the advantages of hierarchical routing with an AODV-based scheme that integrates energy parameter along with the minimum hops count in the decision-making process of data routing. The remainder of the paper is organized as follows: Sect. 2 reviews some related

© Springer Nature Switzerland AG 2020
M. Ezziyyani (Ed.): AI2SD 2019, LNNS 92, pp. 26–37, 2020.
https://doi.org/10.1007/978-3-030-33103-0_4

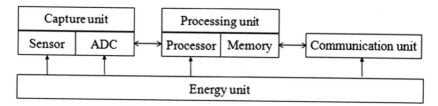

Fig. 1. Composition of sensor nodes.

works, Sect. 3 describes routing in WSNs with a classification of the different routing protocols classes, Sect. 4 discusses the energy model, Sect. 5 presents the proposed routing scheme, Sect. 6 is devoted to simulation environment and experimental results, and finally, a conclusion will summarize the work presented in this paper and discuss several research perspectives related to the proposed contribution.

2   Related Work

Several routing strategies have been suggested in the literature, most of the available research and literature proposing are based on the extension of existing routing protocol to cope with the unreliability of the wireless medium and the energy limitation. In [2], authors proposed a study on how packet loss in the AODV protocol can be minimized in a given network. They developed a technique that identifies the broken link between any two nodes and then repairs the same track. They showed that the simulation results obtained are better than other versions of the AODV protocol. A non-linear dynamic optimization during the AODV route discovery phase by analyzing the packet delivery rate was proposed in [3], the results revealed an improvement of node's data transmission capacity while reducing the packet loss rate. In addition, topology control has been widely studied in order to reduce interference and network energy consumption, in line with this, the work of [4] proposes a solution to reduce the energy cost of packet transmissions in WSNs where sensor nodes create a distributed topology so that their direct neighbors can be reached with minimal transmission power. Other approaches have focused on node power control, allowing their transmission power to be modified to reduce their energy consumption when sending data. Another area of research focused on the notion of the "Duty Cycle" which illustrates this paradigm by allowing nodes to periodically switch from an active mode to a sleep mode. Indeed, the more time a node spends in this state, the more its average consumption decreases, but it is less responsive to network requests (topological changes, data transmissions, etc.). Finally, aggregation is a technique sometimes used to improve the energy efficiency of networks, but is often redefined and/or combined with other concepts to meet the objectives of the algorithms in which it is used.

To the best of our knowledge, our proposed scheme is the first that tries to combine both the advantages of the reactive routing protocol AODV with those of network hierarchization to achieve energy efficiency and optimal data routing.

3 Routing in WSNs

3.1 Classification of Routing Protocols in WSNs

WSNs can be designed under two main topologies:

- Flat topology: All the nodes are homogeneous and identical in terms of capacity and characteristics except the sink, which is responsible for the transfer of data collected to the end user. This topology allows high fault tolerance but it suffers from low scalability.
- Hierarchical topology: In this topology, nodes are divided into several levels of organization and responsibility. Clustering represents one of the most used methods; it aims to divide the network into clusters composed of a Cluster Head (CH) and its cluster members that transfer their collected data for aggregation and transmission to the base station (BTS). This topology increases the scalability of the system, but it causes Cluster Heads overload and an unbalance in the energy consumption on the network.

Several routing approaches have been proposed in the literature with the main objective to achieve good data transmission and optimal use of network resources. The main routing protocols can be classified according to road discovery process, communication and maintenance strategies as shown in (Fig. 2):

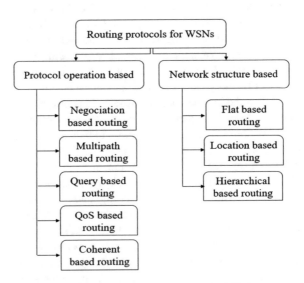

Fig. 2. Routing protocols classes in WSNs.

The instability of the wireless communication medium, the limitation of energy and bandwidth introduce more difficulty and complexity to the design of routing protocols for WSNs. Thus, new routing approaches are needed to perform secure and efficient data routing.

3.2 Overview of the Ad Hoc On-Demand Distance Vector Routing Protocol (AODV)

AODV is a reactive routing protocol that uses four control type of messages: (i) routing request message (RREQ) for message broadcast to another node, (ii) routing reply message (RREP) for the message reception, (iii) routing error message (REPP) for link failure notification and (iv) HELLO message for the links evaluation and detection. In the AODV routing protocol, when a source node cannot find a corresponding entry in the routing table, it may initial a routing discovery procedure by flooding a routing request (RREQ). It also uses sequence number principles to avoid the generation of loops and unnecessary message transmissions on the network, in addition it allows to maintain the consistency of routing information and to use the freshest routes (Fig. 3).

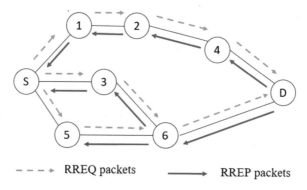

- - - ▶ RREQ packets   ──────▶ RREP packets

Fig. 3. AODV routing mechanism.

In AODV, each node has a routing table that gives information about its neighbors and helps choosing a neighbor that can act as relay for packets transmission. When the source has data to send to a destination, it broadcasts a Route Request (RREQ). When the node receives RREQ, it updates its information for the source node and adds a valid new route to its routing table to reach the source that sent RREQ. When RREQ arrives at the destination, it generates a Route Reply (RREP) response which will be sent back to the source node (Tables 1 and 2).

Table 1. General format of the RREQ packet.

@ Source	Sequence number of the source	Broadcast ID	@ Destination	Sequence number of the destination	Hop count

Table 2. General format of the RREP packet.

@ Source	@ Destination	Sequence number of the destination	Hop count	Life time

4   Energy Model

In WSNs context, energy models are used to evaluate energy consumption of all the network elements and estimate its variation under several deployment conditions, these models can be of three types:

- Analytic energy models that use mathematical description of the operating environment.
- Experimental energy models based on some experimentally measured values of real world devices.
- Theoretical energy models that provide the relation between energy consumption and traffic load in an effective manner.

To model communications between nodes, we used the Unit Disc Graph model, with RT_x as the transmission radius where the network is considered as un-directed graph $G = (E, V)$, it is assumed that two nodes can communicate with each other if the Euclidean distance that separates them is not greater than a given transmission range.

$$E = \{(u, v) \in V^2 | dist(u, v) \leq R_{Tx}\} \tag{1}$$

The energy consumed by the detection modules as well as that of the processing are not taken into account because they are relatively negligible compared to the energy consumed by the communication module. to simplify, we consider that (Fig. 4):

$$E_{tot}(Sensor) = E_{sens} + E_{proc} + E_{com} \tag{2}$$

$$E_{tot}(Sensor) \approx E_{com} \tag{3}$$

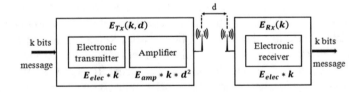

Fig. 4. WSNs energy consumption module [5].

To transmit a message of k bits to a receiver far from d meters, the transmitter consumes:

$$E_{Tx}(k, d) = (E_{elec} * k) + (E_{amp} * k * d^2) \tag{4}$$

To receive a message of k bits the receiver consumes:

$$E_{Rx\,elec}(k) = E_{elec} * k \tag{5}$$

Where E_{elec} represents the electronic transmission energy and E_{amp} the amplifier energy.

5 The Proposed Routing Scheme

Our proposed routing algorithm is an improvement of the AODV routing protocol based on a hierarchization of the network in k layers and on the use of a metric combining energy and distances between nodes in data routing process. Multi-hop routing between layers is provided by the relay nodes responsible also of data aggregation from the lower layer to the upper one which reduces the amount of data and packet redundancy and improves energy efficiency and load balancing. The energy parameter is integrated in routing tables and in the packets route request (RREQ) and route response (RREP) by adding a field containing residual energies of nodes crossed by the route. The proposed routing scheme can be divided into three phases: Initialization by setting-up the k-layers, communication and route maintenance.

5.1 Initialization Phase

The BTS broadcasts a flooding message to calculate/estimate the distance to the nodes at 1-hop which in turn will send this message to the next hop sensor nodes. This operation will be repeated until all the sensors are explored, in the meanwhile, reverse paths to the BTS are established gradually. A layered topology is applied the network nodes according to the number of hops that each of them takes to reach the base station. A layer i represents the sensor nodes with a hop count of 2^i with i e N*. Consequently, several layers are set up with a distance of two hops between a layer N and the layer N−1. This method minimizes the number of control messages exchanged to structure the topology and better optimizes the routing phase. At the end of this phase, each node knows its identifier and layer number reflecting its distance from the base station (Fig. 5).

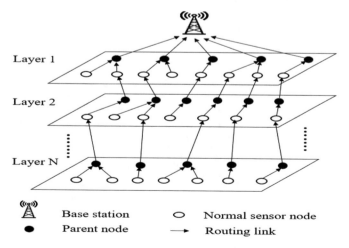

Fig. 5. The layered hierarchization scheme in the proposed approach.

5.2   Communication Phase

When a node has to transmit a route request, it first consults its routing table if an entry exists for that destination. If there is no entry for this specific destination, a route discovery procedure is initiated by a RREQ packet which will be forwarded by every intermediate node to the next node in its layer before it is redirected to the higher layer in the hierarchy until it reaches its destination. Each time the RREQ packet is received by an intermediate node, the cost of the route is calculated and accumulated during the transfer of the RREQ until it reaches the destination. A field called Min_Energy_Cons has been added to the RREQ and RREP packets format to store the minimum value of the energy consumed by the corresponding route, this parameter will be combined with the minimum hop count parameter to calculate routes cost. Each time the RREQ packet is received by an intermediate node, the cost of the route is calculated and accumulated during the transfer of the RREQ until it reaches the destination according to the following formula:

$$\text{Cost(Route)} = \text{Min}\left[\alpha * \sum_{S}^{D} (Energy_{cons}) + \beta * (HopCount)\right] \tag{6}$$

Where A and β are weighting factor that meet $\alpha + \beta = 1$.

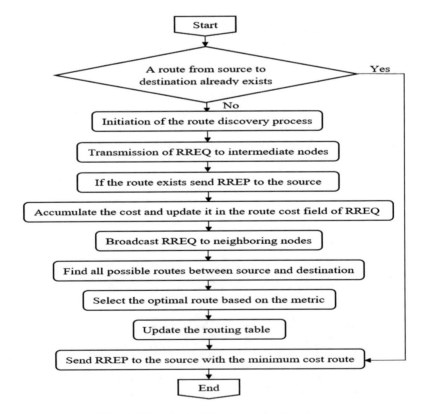

Fig. 6. Flow chart of the communication phase.

The destination chooses then the route with the minimal cost and sends a confirmation message throughout the route. The connection is established upon receipt of this message by the source and communication can begin (Fig. 6).

If after a certain time, no message is received from a neighboring node, the link in question is considered faulty. Then, a RERR message (ERROR Route) propagates to the source and all intermediate nodes will mark the route as invalid and after a while, the corresponding entry is deleted from their routing table.

5.3   Route Maintenance

The path connecting the source and destination through the active entries in the routing tables is called an active path. In case of error, route maintenance can be summarized in the following three points:

- Periodic HELLO messages to detect link breaks.
- Routing tables update based on the sequence number.
- If an intermediate node falls, a special RREP is sent to the source to reset the route discovery process.

HELLO messages are used to maintain connectivity information. At each time interval (hello_interval), the node checks whether it has broadcast a message and if not, it broadcasts a HELLO message to inform neighbors that it is still within radio range. When receiving this message, the node checks whether it has an entry in its routing table to update it; otherwise it adds this node to the list of its neighboring nodes and thus adds the corresponding entry in the routing table.

6   Simulation Environment and Results Analysis

In this part, we compare the proposed routing protocol with AODV (Ad hoc On Demand Distance Vector protocol) and LCH (Layered Clustering Hierarchy Protocol) based on evaluating metrics of energy consumption, end-to-end delay and packet delivery ratio. Our simulations have been run in NS2, results were reproduced in graphs based on trace files generated during simulations. Table 3 lists the simulation parameters.

Table 3. Simulation parameters.

Parameters	Values
Routing protocols	Our protocol, LCH, AODV
Network size	500 m * 500 m
Number of sensor nodes	20–200
Initial energy of sensor nodes	6 J
Packet size	500 bytes
Transmission range	25 m
Mac protocol	802.11, 3 Mbps
α, β	0.5
E_{elec}	50 nJ/bit
E_{amp}	100 pJ/bit/m^2
Simulation time	60 min
Transport protocol	TCP

- Dissipated energy:

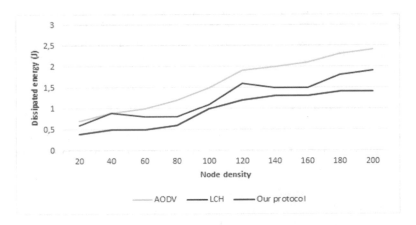

Fig. 7. Evaluation of the dissipated energy.

It can be seen also that, for all protocols, energy decreases proportionally with the increase of the number of nodes deployed in the network. In addition, our protocol offers better energy conservation as a result of the route selection process and the network hierarchization into layers, this approach minimizes communication related to saving the overall network view since each node is only concerned by the structure of the layer to which it belongs (Fig. 7).

- End-to-end delay: The time between transmission and reception of a packet, the lower the value of this metric, the better the performance of the routing protocol.

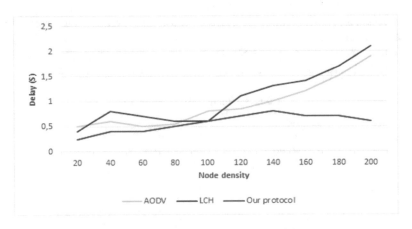

Fig. 8. Evaluation of the end-to-end delay.

As shown in (Fig. 8) the layered network organization reduces the delay since data is sent from a lower layer to a higher layer in the hierarchy as it approaches the Sink. Thus, it distributes routes over the network unlike the AODV protocol which can concentrate routes around a few nodes since it's based only on the minimum number of hops in route finding procedure.

AODV protocol suffers from some link breaks due to congestion, which implies the diffusion of several control packets used for road repair and maintenance resulting in increasing energy consumption. Whereas the LCH protocol does not guarantee a homogeneous distribution of CHs on the network because it is based on the execution of a distributed LEACH like protocol in each layer without taking into consideration the energy of sensors in the CH election process, moreover, distant cluster-heads communicate directly with the sink node which quickly depletes nodes energy in large-scale networks.

- Packet Delivery Ratio: The ratio between the amount of incoming data packets sent by the source node and received data packets at the destination node. This metric implies the accuracy and completeness of the routing technique. The performance is better when this ratio is high.

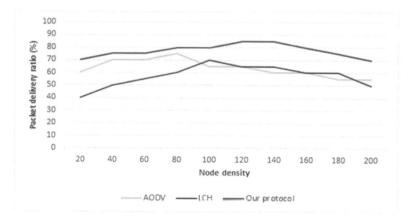

Fig. 9. Evaluation of packet delivery ratio.

Our protocol has better route distribution rather than that of both AODV and LCH protocols, which has a positive impact of the PDR ratio (Fig. 9).

It is shown by simulation that the metric proposed for route selection process offers better performance as it always chooses the shortest paths with minimum energy consumption, this has the effect of prolonging network lifetime and improving packet delivery ratio by reducing the probability of link breaks, also, the organization of the network in layers limits the generation of RREP packets by many intermediate nodes which distributes routes very well and reduces the network load. To sum up, the results obtained show the effectiveness of the proposed routing approach that benefits from two key designs: network hierarchization in layers and the metric used in route discovery process that includes energy parameter along with the minimum hops. This helps to control the transmission radius and requires less transmission power.

7 Conclusion

Wireless sensor networks have become a key technology in the field of intelligent environments. The development of these networks has been accompanied by new challenges particularly due to their limited computing and storage capacities and, most importantly, to their limited energy resource. In this work, we proposed a new routing approach that extends the basic AODV protocol by adopting a metric that combines energy and the number of hops in the route discovery process along with the hierarchization of the network in several layers. Simulation results demonstrated better resource optimization and load balancing improvement. On the basis of the promising findings presented in this paper, work on the remaining issues is continuing and will be presented in future papers.

References

1. Aadri, A., Idrissi, N.: An advanced comparative study of MANETs routing protocols under varied number of nodes and mobility rate. J. Commun. **13**(6), 284–292 (2018). https://doi.org/10.12720/jcm.13.6.284-292
2. Garg, H.K., Gupta, P.C.: Minimization of average delay, routing load and packet loss rate in AODV routing protocol. Int. J. Comput. Appl. **44**, 14–17 (2012)
3. Yuanzhou, L., Weihua, H.: Optimization strategy for mobile ad hoc network based on AODV routing protocol. In: The 6th International Conference on Wireless Communications Networking and Mobile Computing (WiCOM), Chengdu, pp. 1–4 (2010). https://doi.org/10.1109/wicom.2010.5601193
4. Abd Aziz, A., Sekercioglu, Y.A., Fitzpatrick, P., Ivanovich, M.: A survey on distributed topology control techniques for extending the lifetime of battery powered wireless sensor networks. IEEE Commun. Surv. Tutorials **15**(1), 121–144 (2013)
5. Heinzelman, W.R., Chandrakasan, A., Balakrishnan, H.: Energy-efficient communication protocol for wireless microsensor networks. In: The 33rd Annual Hawaii International Conference on System Sciences (HICSS 2000), Washington, DC, USA (2000)
6. Tawalbeh, L., Hashish, S., Tawalbeh, H.: Quality of service requirements and challenges in generic WSN infrastructures. In: The International Workshop on Smart Cities Systems Engineering (SCE 2017), vol. 109, pp. 116–121 (2017). https://doi.org/10.1016/j.procs.2017.05.441.1
7. Chaudhary, D.D., Waghmare, L.M.: Energy efficiency and latency improving protocol for wireless sensor networks. In: The International Conference on Advances in Computing, Communications and Informatics (ICACCI), Mysore, 2013, pp. 1303–1308 (2013). https://doi.org/10.1109/icacci.2013.6637366
8. Gupta, V., Pandey, R.: An improved energy aware distributed unequal clustering protocol for heterogeneous wireless sensor networks. Eng. Sci. Technol. Int. J. **19**(2), 1050–1058 (2016). https://doi.org/10.1016/j.jestch.2015.12.015

9. Abarna, S., Arumugam, N.: Performance comparison of various energy aware protocols for WSN applications. Int. J. Innovative Res. Electron. Commun. 1(9), 1–8 (2016)
10. Touati, Y., Ali-Chérif, A., Daachi, B.: Optimization techniques for energy consumption in WSNs. In: Energy Management in Wireless Sensor Networks, pp. 9–22 (2017). https://doi.org/10.1016/B978-1-78548-219-9.50002-3
11. Shen, H., Bai, G.: Routing in wireless multimedia sensor networks: a survey and challenges ahead. J. Netw. Comput. Appl. 71, 30–49 (2016)
12. Bajaber, F., Awan, I.: Adaptive decentralized reclustering protocol for wireless sensor networks. J. Comput. Syst. Sci. 77, 282–292 (2011)
13. Adam, M.S., Hassan, R.: Delay aware reactive routing protocols for QoS in MANETs: a review. J. Appl. Res. Technol. 11(6), 844–850 (2013)
14. Channappagoudar, M.B., Venkataram, P.: Performance evaluation of mobile agent based resource management protocol for MANETs original research article. Ad Hoc Netw. 36(1), 308–320 (2016)
15. Elappila, M., Chinara, S., Parhi, D.R.: Survivable path routing in WSN for IoT applications. Pervasive Mobile Comput. 43, 49–63 (2018)
16. Kavi Priya, S., Revathi, T., Muneeswaran, K., Vijayalakshmi, K.: Heuristic routing with bandwidth and energy constraints in sensor networks. Appl. Soft Comput. 29, 12–25 (2015)
17. Singh, R., Verma, A.K.: Energy efficient cross layer based adaptive threshold routing protocol for WSN. AEU- Int. J. Electron. Commun. 72, 166–173 (2017)

Analysis and Evaluation of Cooperative Trust Models in Ad Hoc Networks: Application to OLSR Routing Protocol

Fatima Lakrami[1(✉)], Najib EL Kamoun[1], Ouidad Labouidya[1], and Khalid Zine-Dine[2]

[1] STIC Laboratory, Faculty of Sciences, Chouaib Doukkali University, El Jadida, Morocco
{lakrami.f,elkamoun.n,laboudiya.o}@ucd.ac.ma
[2] LAROSERI Lab, Faculty of Sciences, Chouaib Doukkali University, El Jadida, Morocco
zinedine@ucd.ac.ma

Abstract. Mobile Ad Hoc Networks are wireless networks, of which the installation does not require any existing infrastructure, this property has induced the philosophy of "all distributed" and distributed control. The departure and arrival of a node is autonomous and does not interfere with the continuity of services. These advantages allow the deployability of these networks in situations where the use of wired networks is expensive, difficult or even impossible. However, with these advantages, some security concerns are emerging. Indeed, with a shared communication medium and in the absence of a central entity or a fixed infrastructure, traditional security solutions are not adapted, it is now difficult to manage encryption keys, certificate distribution, and trust management between nodes. As a node does not necessarily have knowledge about others when the network is growing, trust a priori may not exist. We are interested in this paper to study security problems in Manets in general, and more particularly those related to trust management. This paper detail, the principle, the constraints, and the challenges of deploying suitable trust models, to deal with Manets constraints. An example of a light trust model for enforcing security in OLSR routing protocol used for Manets, is presented and evaluated.

Keywords: Ad Hoc network · Distributed control · Selfishness · Security · Trust models · OLSR

1 Introduction

Implementing a solution to ensure that mobile adhoc networks (MANETs) to be secure is not a simple task. In fact, adhoc mobile networks are faced with many problems related to their features that make security solutions developed for wired or wireless networks with infrastructure, inapplicable in the context of mobile adhoc networks. Among the vulnerabilities that affect adhoc mobile networks there are: The lack of infrastructure, the dynamic network topology, the vulnerability of the nodes, the shared

© Springer Nature Switzerland AG 2020
M. Ezziyyani (Ed.): AI2SD 2019, LNNS 92, pp. 38–48, 2020.
https://doi.org/10.1007/978-3-030-33103-0_5

channel, and limited resources. Such problems make Manets more exposed to certain types of attacks than wired networks, especially while considering a distributed and random architecture.

In another context, there is routing. In fact, and Unlike wired networks, where dedicated nodes are used to support basic functions like packet forwarding, routing, and network management, in adhoc networks, those functions must be ensured out by all available nodes. This particularity is at the core of the security problems that are specific to adhoc networks. In fact, the nodes of an adhoc network cannot be automatically trusted to execute critical network functions such as routing or packet forwarding. The solution is mainly to put in place a number of mechanisms allowing a self-administration and a management of self-organized nodes by designing new settlement protocols and trust management system including security protocols for these mobile entities.

Trust is a fundamental psychological and cognitive concept on which all types of transactions and communications are based. In an open adhoc network, a node relies on nodes cooperation if it has some information to exchange. This assumes that the node needs to give his trust to other nodes and to their ability to forward information to the targeted destination. A priori, a trust relation can only exist by default in a few special scenarios like military networks requiring specific hardware for the implementation of such critical functions, in such scenarios, and for a large network, an entity authentication raises key management requirements.

Indeed, trust management relies on a number of fundamental concepts, the most commonly used is cryptography. Several security architectures have been proposed to distribute the encryption keys as well as the certificates for the purpose of securing communication between the nodes. However, most of these architectures do not respect characteristics of adhoc mobile networks, and do not meet several new requirements, such networks who shares the same characteristics as social networks models, and then there are keys, that need to be predetermined by a third party agent/server, wish increases the computational overhead significantly leading to performance degradation. To overcome such problems, enforcing security through seamless trust models attains significant interest to the research community. The deployment of any security service requires the definition of a trust model that defines who trusts who and how. Recent research invested considerable efforts in enhancing trust models framework to securing mobile adhoc networks in a more efficient way. In this paper, we mainly focus on trust approaches based on Cooperation enforcement. We will review their principle, while investigating some developed models, used to increase security levels in Manets. An example of implementing a cooperative trust model in OLSR will be introduced.

The rest of the paper is organized as follows: In Sect. 2, a review of security in Manets is presented, considering a classification of attacks and security solutions. Section 3 focuses on describing trust models in Manets. In Sect. 4, we develop a simple approach for enhancing security in OLSR routing protocol based on implementing a trust model for relaying traffic. Section 5 concludes the paper and highlight some perspectives.

2   Review of Security in Manets

In a wireless environment, several attacks are exploitable. In plus of identity usurpation, there are many other attacks that exploit the shared physical medium to gain a free access to the network, which gives any machine the opportunity to get into the coverage distance of other machines on the network and thus to perform attacks that can affect the proper functioning of the network. Also at the MAC level, and by scanning the access mechanisms used by the physical and MAC layers in an adhoc network, a malicious node with sufficient transmission power can then prevent the other nodes from accessing the communication channel, for a certain period of time, by acquiring the channel as long as possible through the modification of MAC transmission properties. At the network layer, a hop by hop routing is performed, which means that any machine that infiltrates the network will have the opportunity to intercept and route packets to (the right/bad) destination, or even refuse to relay packets [1, 2].

Attacks on a network come in many varieties and they can be grouped based on different characteristics. So different classification aspect were proposed by the literature. Figure 1 propose a hybrid classification through a taxonomy that combines the three following criteria: the origin of the attack (external, internal), the type of the attack (passive, active) and the Protocol layer of the attack (OSI model). Some authors [3], consider as well, the trustworthiness of communication partner in the network, however, we assume that this parameter is more related to internal attacks, and is considered as a propriety of security solutions.

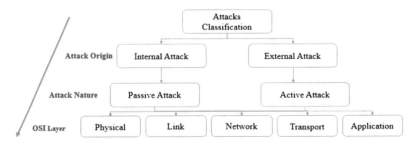

Fig. 1. Generic classification of attacks in Manets

A malicious node infiltrating the network will have the opportunity to modify, usurp, inject data and even generate false messages. This kind of behavior can impact the cooperation of nodes necessary for the establishment of a network without infrastructure. An attack can take the form of "selfishness", and this occurs when one or more nodes refuse to relay traffic to the rest of the network nodes (routing), either to preserve its energy or for malicious intentions, which can implicitly prevent communication between the nodes. Figures 2 and 3 express the evolution of delay and throughput according to the increase of the number of "selfish" nodes in an adhoc network composed of 100 nodes, simulated on NS2, and using OLSR as routing protocol.

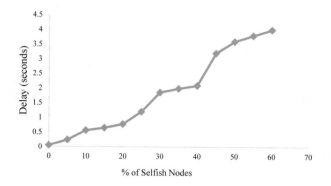

Fig. 2. Evolution of delay (sec) in function of % of selfish nodes

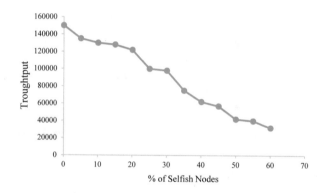

Fig. 3. Evolution of the throughput in function of % of selfish nodes

Another problem, as crucial as those mentioned above, related to the fact that nodes in the Ad-hoc mobile network can join and leave the network freely without notice. Thus it is difficult in most cases to have a clear picture of the distribution of nodes in the network. In such a context, the notion of trust between nodes in an adhoc network becomes a crucial parameter in the functionality and continuity of the network. This instantaneous joining and departure causes an evolution of trust relationships that may change over time [4].

Security in adhoc networks, as well as availability, reliability… is an essential characteristic, at least to a minimum degree, in all applications, some of which may have higher requirements (e.g. military), which means that the protocols used in adhoc networks must be resistant to a large number of attacks, to better detect their failures. However, no final overall solution has been developed so far. Indeed, the modularity of the security, and specifically in a very vulnerable architecture such as that of the Manets, makes this task very difficult. However, it is very important to be able to locate the weak point of the mesh in an adhoc network, is that of the access to the network, the establishment of a trust relation, the centralization or the distribution of the control, the security of network traffic.

In support of this, research on risk and attack proposes multiple formal and semi-formal methods for modeling and simulating attacks and risks in Manets. In general, it covers the 4 aspects described by Fig. 4. the deployment of a security system varies according to its final destination [5]: detection (IDS) or prevention (IPS) of an attack. The best scenario would be to be able to protect the network from intrusions, or, if so, to be able to detect it in time. in general, in wireless networks, and due to the nature of the transmission medium, it is difficult to prevent access to traffic in its native format. However, it is possible to install a security system to manage trust between the nodes belonging to the network, and succeed to locate an intrusion from a malicious node and stop it [6, 7].

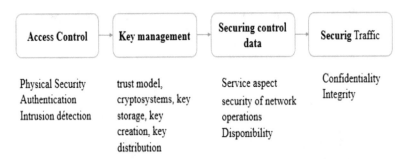

Fig. 4. Basic security solution developed for adhoc networks context

3   Trust Models in Manets and Challenges

Security in adhoc networks, as well as availability, reliability… is an essential characteristic, at least to a minimum degree, in all applications, some of which may have higher requirements (e.g. honeycomb), which means that the protocols used in adhoc networks must be resistant to a large number of attacks, to better detect their failures. Security solutions must offer certain basic services such as: authentication, integrity control, confidentiality, availability and non-repudiation. The majority of security solutions proposed by the literature are based on symmetric or asymmetric cryptography. But the major problem with these solutions in the environment of mobile adhoc networks is the management and distribution of encryption keys. Proposing a single certification authority (CA) for the entire network is not a desirable solution because this design is vulnerable to type attacks (DoS) [8], some solutions propose hierarchical (clustering) or group-based management architectures, consisting on electing an entity (with the highest degree of trust) as a group controller. We will explain in the following the principle of the trust model, and the main elements for building and maintaining it.

Recently, several IDS developments have relied on trust models associated with machine learning systems to generate more correct and effectively unbiased outputs. The performance of such systems explicitly depends on the accuracy and credibility of the previously established trust model [9].

Building trust in adhoc network can be established through giving nodes the possibility to check the confidence level of their neighbors being trustworthy and secure, this should invoke the deployment of signatures and encryption mechanisms.

In Manets, trust models are designed based on a number of criteria, such as perceptions, metrics and properties of trust [10], and as illustrated by Fig. 5 the relationship of trust is based during its establishment on classical cryptographic mechanisms, however the contexts proposed so far, are not generic, and cannot be extended to all situations.

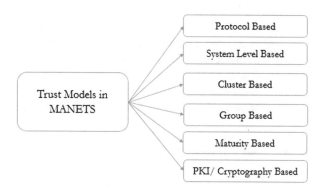

Fig. 5. Classification of trust models in Manets

When it comes to trust, several problems can arise:

- How to estimate the degree of trust of a node that has just joined the network, would it be wise to assign it a high or low trust value?
- How to designate the entity responsible for authenticating and registering a new node, if we consider a trust model based on the PKI?
- How to monitor the evolution of the confidence level and which parameter should be selected?
- How can a single "eligible" entity manage a network despite its extension?

The majority of the problems or inter-network attacks in the Manets result directly from the non-cooperation between the different entities forming the network. This problem has gained a lot of attention in the context of adhoc network management, and more techniques have been proposed to mitigate this type of threat.

Trust management in adhoc networks is divided into three broad categories of models: Reputation and Recommendation-based trust models based on certification (credentials) and systems developed from the social network of the user.

In the first category, trust is based on the concept of reputation. The reputation of a node increases when it correctly performs the tasks that correspond to the correct network operation, such as routing. Each node observes the behavior of neighbors and declares an accusation if he considers that a node is suspect, which allows to isolate all the malicious nodes. In the second category, trust management is essentially through the intervention of a certification authority which can be central or distributed.

The role of this authority is to check whether a given node is trustworthy or not, in order to give it a certificate, which is a data structure in which a key is linked to an identity, in order to prove its legitimacy. The main purpose of such systems is to allow access control. As a result, their concept of trust management is limited to the policy rules defined by each application of the network, whose provision of the requested service is validated only after validation of the certificate of the node requesting the service. In the third category, systems analyze the social network of each node, namely its existing relationships in order to derive conclusions about the levels of trust to be given to other nodes, they are based on mechanisms of reputation, credibility, of honesty and also processes of recommendations.

In Cooperative trust models are based basically on reputation among nodes. The reputation of a node increases when it carries out correctly the tasks of route construction and data forwarding. The models of this category support effective mechanisms to measure the reputation of other nodes of the network. They also incorporate techniques that isolate the misbehaving nodes that are those that show a low reputation value. Trust models based on cooperation enforcement are well surveyed in the literature [11, 12].

4 Proposing a Trust Model for Manets at Routing Level: OLSR

The recent security enforcement mechanisms in MANET are based on evaluating trust and trustworthiness of the nodes or agents participating in communication. The trust is defined as a quantitative value, needed to be calculated by all mobile nodes in order to authenticate each other, it is considered (in human language) as the degree of belief that a node acts as expected in achieving the requiring tasks to ensure connectivity and continuity of an ad hoc network services.

In a routing process, neighbor node's distrust is evaluated by the sender by observing activities carried out by that neighbor. To be specific, a node ni will increase the distrust score of its neighbor nj if the nj does not forward the packet sent by ni. Trust could be defined as a risk factor, or as a belief Trust degree, also as subjective probability [13].

We propose in this section to develop a simple mechanism for integrating trust in Manets at routing level, we choose to work, for this purpose, with OLSR routing protocol.

OLSR (optimized link state protocol) is developed for mobile adhoc networks. It operates as a table driven and proactive protocol, thus exchanges topology information with other nodes of the network regularly. The nodes which are selected as a multipoint relay (MPR) by some neighbor nodes announce this information periodically in their control messages. Thereby, a node announces to the network, that it has reachability to the nodes which have selected it as MPR [14].

In route calculation, the MPRs are used to form the route from a given node to any destination in the network. The protocol uses the MPRs to facilitate efficient flooding of control messages in the network, thus the MPRs play the most important role in the functioning of the protocol, they are responsible on relaying broadcast traffic and also to form route from any source to any destination in the network.

The willingness is a parameter that defines the ability of the node to become an MPR. The willingness of a node may be set to any integer value from 0 to 7, and specifies how willing a node is to be forwarding traffic on behalf of other nodes. Nodes will, by default, have a willingness WILL_DEFAULT. WILL_NEVER indicates a node which does not wish to carry traffic for other nodes, for example due to resource constraints (like being low on battery).

In our approach, we focus our evaluation on nodes eligibility to become MPRs, the proposed model aims in a first way to resolve the problem of selfishness. In this context, nodes must express their ability to become MPR, based on their residual energy and the percentage of reachability of 2-hop neighbors. We define for this purpose the following parameters:

Trustworthiness: is objective probability by which the trust performs a given action on which the welfare of the trustor depends, in our case, we link it to the willingness, of a node to become MPR, the willingness is calculated in function of the residual energy and the percentage of reachability of 2-hop neighbors.

$$w = \alpha \operatorname{Pr} + \beta \operatorname{Re}$$

Trust: is the subjective probability by which the trustor expects that the trustee performs a given action on which the welfare of the trustor depends. In the proposed model we, evaluate the trust, based on two parameters: the willingness, calculated from the equation above and the number of forwarded packets, considering that, the node subject to trust is 1-hop neighbor from their evaluators. The history of a node is then recorded and enable to extract the probability of the node continuing being an MPR is sustainable or not.

We consider, the trust model explained by [8], where a trust relationship can be represented as T {subject: agent; action}, in this definition, the subject trusts the agent with value T that the agent will perform the action. The probability that the agent will perform the action in the subject's point of view can be represented by P {subject: agent; action}. An OLSR node i will trust a node j to become an MPR if the conditions above. Further information are presented by [8].

The new version of OLSR named SOLSRv1, it represents a test version of implementing a trust model in the OLSR routing protocol. In this version, the "action" will be either the node can be an MPR, or not. Considering that MPRs plays the most important role in an OLSR topology, so it is important to take wise decision while deciding about it. The proposed model was implemented in NS2 simulator, with 100 nodes, exchanging light CBR traffic. We address our model to deal with the problem of selfishness. And we consider the following metrics to measure the performance of the proposed model:

- End to end delay
- Loss Rate
- No of misrouted packets

As Fig. 6 shows, the first implementation of this model enable to enhance the network performance in term of delay, loss rate and misrouted packets. If fact, a node selfishness, is expressed through a week willingness value, the node will abstain itself

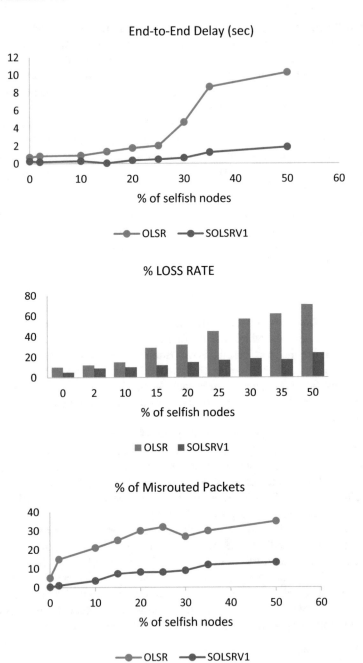

Fig. 6. Performance comparison of OLSR and SOLSRv1 that implement a trust model based on the willingness.

from relaying traffic, and then becoming an MPR. But for the first time the joins the network, and due to its high willingness, the node will be relaying traffic, before the score of trust start to decrease. The proposed solution does not generate any traffic overhead and represent a first simple implementation, to test the efficiency of dealing with selfishness through developing a simple mechanism to take the rightful actions toward selfish nodes.

SOLSRv1 shows better performance than the original OLSR, because in SOLSRv1, selfish nodes won't be considered during MPR selection and routing phasis. This approach enable the network to consider only nodes able to relay traffic, according to their residual energy and the number of covered 2-hop neighbors. If the trust evolves positively (the number of relayed packets is correct), the node is considered as trust-worth, and continue to act as an MPR. Otherwise, the node is considered as selfish, and will be banished from MPR group.

5   Conclusion

Developing trust modeling is a very complex task, this is due to the fact that the elements required to establish trust and estimate reputation often changes according to requirements and context. This can lead to errors in the design and implementation of trust models. Second, a node's trust remains subjective as long as the other nodes generally have different levels of experience with it. Trust is also an incomplete and non-transitory property. This leads to the need to use recommendations from one or more third parties in order to give more precision to the confidence assessment.

Moreover, trust is asymmetric, if we consider resource constraints as an element of evaluation. In fact, different nodes have different levels of resource constraints and, therefore, a higher capacity node may not trust lower capacity nodes. Finally, trust depends on the context, to perform a computationally intensive task, a node with high computational power can be considered reliable, while a node with low computational power, although not malicious, can be suspicious.

It is therefore imperative to think carefully about studying and developing a trust model, which is as generic as possible, with a minimum margin of asymmetry and subjectivity. That being said, existing models currently attempt to design hybrid esti-mations, by weighting the coefficients of the calculation parameters relative to the value or rather the degree of confidence. In future work, we will focus on the study and evaluation of this type of system, while remaining within the framework of routing and specifically OLSR, which is an interesting test model to work on it.

References

1. Gagandeep, A., Kumar, P.: Analysis of different security attacks in MANETs on protocol stack A-review. Int. J. Eng. Adv. Technol. (IJEAT) 1(5), 269–275 (2012)
2. Rajkumar, K., Prasanna, S.: Complete analysis of various attacks in Manet. Int. J. Pure Appl. Math. 119(15), 1721–1727 (2018)

3. Beghriche, A., Bilami, A.: Un modèle de Sécurité basé sur la Confiance Floue Pour Assurer la Qualité de Service (QoS) dans Les Réseaux Mobiles Adhoc
4. Pooja, P., Puja, P.: Trust based security in Manet. Int. J. Res. Eng. Technol. **5**, 11 (2018)
5. Singh, D., Bedi, S.S.: Novel intrusion detection in MANETs based on trust. Int. J. Comput. Sci. Inf. Technol. **6**(4), 3556–3560 (2015)
6. Maglaras, L.A.: A novel distributed intrusion detection system for vehicular adhoc networks. Int. J. Adv. Comput. Sci. Appl. (IJACSA) **6**(4), 101–106 (2015)
7. Shams, E.A., Rizaner, A.: A novel support vector machine based intrusion detection system for mobile adhoc networks. Wireless Netw. **24**(5), 1821–1829 (2018)
8. Tripathy, B.K., Bera, P., Rahman, M.A.: Analysis of trust models in Mobile Adhoc Networks: a simulation based study. In: The 8th International Conference on Communication Systems and Networks (COMSNETS), pp. 1–8. IEEE, January 2016
9. Jhaveri, R.H., Patel, S.J., Jinwala, D.C.: DoS attacks in mobile adhoc networks: a survey. In: Second International Conference on Advanced Computing & Communication Technologies, pp. 535–541. IEEE, January 2012
10. Omar, M., Challal, Y., Bouabdallah, A.: Certification-based trust models in mobile adhoc networks: a survey and taxonomy. J. Netw. Comput. Appl. **35**(1), 268–286 (2012)
11. Cho, J.H., Swami, A., Chen, R.: A survey on trust management for mobile adhoc networks. IEEE Commun. Surv. Tutorials **13**(4), 562–583 (2010)
12. Marias, G.F., Georgiadis, P., Flitzanis, D., Mandalas, K.: Cooperation enforcement schemes for MANETs: a survey. Wireless Commun. Mobile Comput. **6**(3), 319–332 (2006)
13. Mishra, A., Nadkarni, K.M.: Security in wireless adhoc networks. In: The Handbook of Adhoc Wireless Networks, pp. 477–527. CRC press, Boca Raton (2002)
14. Zougagh, H., Toumanari, A., Latif, R., Idboufker, N.: A novel security approach for struggling black hole attack in optimised link state routing protocol. Int. J. Sens. Netw. **18**(1–2), 101–110 (2015)

Assessing NoSql Approaches for Spatial Big Data Management

El Hassane Nassif[1]([✉]), Hajji Hicham[1], Reda Yaagoubi[1],
and Hassan Badir[2]

[1] Ecole ESGIT, IAV H2, Rabat, Morocco
nassif.hassane@gmail.com,
{h.hajji, r.yaagoubi}@iav.ac.ma
[2] Ecole ENSAT Tanger, Tangier, Morocco
hbadir@uae.ac.ma

Abstract. Since the advent of social networks, Iot and Smartphones, spatial data are taken by storm by the Big Data phenomenon. Their management and analysis is a real challenge for traditional geographic information systems. Indeed, these solutions don't respond effectively to big data constraints as they still rely on relational databases to manage and process spatial data.

In this paper, we compare, from a qualitative and quantitative point of view, three families of NoSql databases with geospatial features: Key-value, column and document oriented. We explore the ways offered by these three NoSql paradigms for efficient management and analysis of massive spatial vector data and then we analyze the performance of two of them. The empirical evaluation is performed on two clusters based on an open data datasets and gives some advantages and limitations of these approaches.

Keywords: Spatial Big Data · Distributed spatial computing · Accumulo · Elasticsearch · Redis · Spatial · Vector

1 Introduction

Relational databases have revolutionized data management by structuring data since their appearance fifty years ago. Codd [1] explains that the relational model uses algebra notions to remedy to hierarchical and networked model limitations [2]. In addition and thanks to Sql data query language, relational model has remained the predominant choice for storing and querying structured data [3].

However, the big data phenomenon [4] is changing this situation. Due to the advent of social networks, Iot and Smartphones, spatial data become easy to collect and have acquired Big Data characteristics such as volume, velocity and variety. This situation is a challenge for traditional spatial solutions that uses RDBMS as back-end systems for geospatial services and operations such as PostGIS on PostgreSQL, Oracle Spatial on Oracle. Amiran [5] explains that even if relational databases still can be used in many geospatial data-related tasks, they are not efficient enough to handle geospatial Big

© Springer Nature Switzerland AG 2020
M. Ezziyyani (Ed.): AI2SD 2019, LNNS 92, pp. 49–58, 2020.
https://doi.org/10.1007/978-3-030-33103-0_6

Data. In the last years we noted the emergence of new data store models called NoSql[1]. They provide a solution to relational database users' communities in order to overcome constraints and challenges related to big data [6]. It's already established that NoSql database exceeds relational databases in terms of performance, elasticity and agility. This comparison has already been made on the occasion of several research works. Mukherje [7] made a comparison work between RDBMS and NoSql and concludes in his paper that NoSql are the response to the limitations of relational databases in the way that they can deliver performance, scalability, and flexibility. When it comes to the spatial domain, Laksono [16] compared Mongo DB to PostGIS, his results show that NoSql document oriented database, i.e. MongoDB, performs better in loading big geospatial data compared to traditional SQL database using PostGIS. Baralis [17] compared Azure SQL database to Azure document Database, results shows that the document oriented database gives the least average response time with respect to the number of concurrent users.

In this paper, instead of focusing our research efforts on comparing NoSql to relational databases we followed a different approach. We set ourselves the goal of exploring and comparing spatial features offered by three types of NoSql databases: key-value, document, and column oriented. We also compare the performance of ingestion processes of each of these databases by calculating the time required to load data into the database. We used several versions of data by increasing the number of rows progressively from thousand records up to one billion. The second comparison we did in our work concerns spatial query performance. The queries involved are: bounding box filter, distance filter, within polygon filter, k nearest neighbours (knn). The document is organized as follows: a first part to introduce NoSql databases and to detail spatial features offered by each of them. A second part describing the material, the data used as well as the method followed and final results. In the end, a synthesis and prospects of our work are given.

2   NoSql Paradigms

The term "NoSql" appeared in 1998 when Strozzi [8] named his open-source database "NoSql" in order to indicate that his solution does not provide any SQL interface even if it is based on a relational database. This has no relation to the recent NoSql movement who defends the use of other models different from the relational paradigm. In fact, if it was necessary to rename this movement it will be given the name "NoRel" for non-relational [8]. Later, the term "NoSql" was coined in 2009 to describe a movement whose ambition is to create new data storage models that deviate from the relational paradigm on several aspects, particularly in terms of data structuring [9].

NoSql databases are not intended to replace relational databases, but they are an alternative, an add-on that brings more interesting solutions. Indeed, they are better adapted to massive data than the relational model and allow managing complex and heterogeneous objects with no pre-defined schema. There is a large variety of NoSql

---

[1] Not only sql.

databases, each meeting specific needs [10]. The purpose of this article is not to develop a complete panorama of existing solutions. However, we group all approaches into three[2] categories that we describe succinctly in the following sections.

2.1 Key Value Oriented Data Store

Key-value databases manage data using large hash tables based on Key-value pairs to store data records. Access to data is granted through keys that act as a search index in the Hash table [11]. The data value can be as simple as a string of Latin letters or can use more complex types. Several key-value databases exist on the market, they each have a non-standardized programming interface for data access.

As part of our work, we select Redis[3], a NoSql database that has the particularity of loading data records in RAM[4] in addition to having a key-value schema. Persistence on the disk is granted by regular snapshots of all in memory loaded data. Scalability is guaranteed by replicas on secondary servers which provide additional capacity if needed. The replication operation is done by performing an initial database load on replica servers and then by sending incremental updates to them.

Concerning spatial data, Redis uses a native geometry Point object to represent a single location in space with an x-coordinate (longitude) and a y-coordinate (latitude). Just like standard data types, storage of spatial data is based on sorted datasets called GeoSet. Sort order is maintained by using a Hash function that transforms the two spatial dimensions into one value named score. It's a double precision number used for sorting and can represent any different decimal or integer value between $-/+2 \wedge 53$. But when representing much larger numbers, elements will be added to the GeoSet with the same score. This problem was solved by implementing a lexicography technique that considers strings as binary data and compares the raw values of their bytes, byte after byte. Spatial data inserts and updates are done using the GeoAdd method, deletes are done by GeoRem method and searching through GeoPos or GeoHash methods. GeoPos return X, Y coordinates, and GeoHash return the linearized result using the hash function previously described. For spatial queries, Redis offers the GeoRadius method that allows you to query the database to render all the points included in a circle with a predefined radius. GeoRadiusByMember does almost the same thing but considering the point entered as the center of the circle in question. Polygon and multi-polygons are not supported but there is an extended library Lua library[5] which provides support for these two structures. It also allows search operations through GeometryGet method, which can be used to retrieve the coordinates of geometry through its key. The result can be enriched with other information depending on the parameters injected. With Perimeter for example allows retrieving the length of the edges in the case of a polygon, WithBox allows to retrieve the geometry bounding box. Also, WithCerle allows

---

[2] Graph-type database family is not included in this paper.

[3] https://redis.io.

[4] Random access memory.

[5] https://github.com/nrk/redis-lua.

retrieving the minimum bounding circle. Finally, and concerning the encoding, Redis allows simple feature GeoJson[6] conversion.

2.2 Document Oriented Data Store

Document oriented data stores are based on the previous key-value paradigm with the particularity that the data value is a document (JSON, XML and many other formats). This way of structuring data allows us to manipulate a set of linked information through a single key, since the entire row data is stored inside a document. This is equivalent to the multi-table join operation we usually use to retrieve data spread across several relational tables. In this paper, we study Elasticsearch[7], an open source document-oriented NoSql database, with distributed search and analytical engine built on Apache Lucene[8]. The distributed nature of Elasticsearch enables it to process large volumes of data in parallel. Inside an Elasticsearch cluster[9], each node houses a single instance and is by default considered as a data node able to store shards[10]. Elasticsearch names one or more nodes as dedicated masters to be responsible for lightweight cluster-wide actions such as creating or deleting an index, tracking which nodes are parts of the cluster, and deciding which shards to allocate to which nodes.

Elasticsearch stores data in the form of JSON documents and when a new document is inserted, it automatically stores the original value and adds a searchable reference to be used later for searching through Elasticsearch API. Storing data in Elasticsearch is called indexing, but before we can index a document we need to choose the storage type mapping. A type mapping represents a class of documents that have a similar structure. The type is defined through its name and the attributes of the documents as well as their corresponding types.

As for spatial vector data, Elasticsearch supports two types: geo_point (supports latitude and longitude pairs) and geo_shape (supports Point, MultiPoint, LineString, MultiLineString, Polygon with holes and MultiPolygon with holes). Their implementation follows OGC[11] and ISO standard specifications[12]. Spatial indexing is done through Geohash base32 encoded strings of the latitude and longitude and a Bkd-Tree index [12]. Longitude/latitude coordinates default size is 16 Bytes × 2. When indexed as Geohash, they will be base32 encoded strings. The default precision is ~2 cm base on 12 characters in the encoded hash string, each character in a GeoHash adds additional 5 bits to the precision. So the longer the hash string is, the more precise the location will be. Because of memory allocation, precision can be a bottleneck for Elasticsearch. Depending on the use case, the precision should be handled carefully

---

[6] http://geojson.org/.

[7] https://www.elastic.co/fr/.

[8] https://lucene.apache.org/core/.

[9] A logical and physical collection of nodes.

[10] A part of an index distributed, automatically managed and replicated across the Elasticsearch cluster.

[11] https://www.opengeospatial.org/standards.

[12] https://www.iso.org/committee/54904/x/catalogue/.

by specifying how precise we need our point to be instead of the default full 12 levels of precision.

Elasticsearch offers several spatial queries: Geo_Shape query that finds documents with geo-shapes which either intersect, are contained by, or do not intersect with the specified Geo-Shape. Geo_Bounding_Box query finds documents with Geo-Points that fall into the specified rectangle. Geo_Distance query, finds documents with Geo-Points within the specified distance of a central point. Geo_Polygon query Find documents with Geo-Points within the specified polygon. Elasticsearch use Kibana[13] as a geotool to visualize spatial data on the map and to interact with it: Zoom In/Out, Drill Down, Filter and also allows heat map creation based on data layers built on base map tiles. Elasticsearch can be used as GeoServer Data Store[14] which allows consuming geospatial features from an Elasticsearch index. OGC filters are converted to Elasticsearch queries and can be combined with native Elasticsearch queries.

2.3   Column Family Oriented Data Store

Column-oriented databases share a few terms with relational databases such as records or columns. However there is still a huge difference between these two models [13]. The number of columns can vary from one record to another. This approach does not require pre-defined or fixed schemas. The column is thus considered as a storage unit with a name and a value. Each record has a unique key and a set of columns or attributes. It is not uncommon for this type of database to support millions of columns [14]. For this reason columns are grouped into collections called column families in order to facilitate storage, organization, and partitioning of data.

As part of this work we selected Accumulo[15] which is one of the many implementations based on Google BigTable [14], highly scalable, high-concurrency, high-reliability and fault-tolerant distributed column-oriented data store. Accumulo is built on top of the Hadoop Distributed File System[16], providing powerful storage capacity and low-latency query service. At the most elementary level, data is stored according to a key-value schema and follows a column family data model. Access to data therefore is granted by a composed key: RowId, a column, and a time stamp. Unlike relational databases, IDs are not sequentially generated by the data store Accumulo but it is up to the application to handle this task. The ID is used to group all the key-value pairs that have the same value, these pairs are sorted in this order: rowid/column/timestamp. Accumulo is not a spatial database and does not know how to represent spatial objects such as points, lines, or polygons. It also does not know how to perform spatial indexing, spatial queries and spatial operations. But there is a suite of tools called GeoMesa[17] built on top of Accumulo. And uses it as a distributed NoSql data store with a set of tools for indexing, managing and analysing spatial data. For indexing,

---

[13] https://www.elastic.co/fr/products/kibana.

[14] https://github.com/ngageoint/elasticgeo.

[15] https://accumulo.apache.org/.

[16] https://hadoop.apache.org/docs/stable/hadoop-project-dist/hadoop-hdfs/HdfsDesign.html.

[17] https://www.geomesa.org/.

GeoMesa use's space-filling zcurve [15] technique to map multi-dimensional data to a single lexicographic value that can be stored in accumulo. In our case we will use two kind of zcurve indexes: Z2 and Z3. The Z2 index uses a two-dimensional Z-order curve to index latitude and longitude for point data. This index will be created if the feature type has the geometry type Point. This is used to efficiently execute queries with a spatial component but no temporal component. Z3 index uses a three-dimensional Z-order curve to index latitude, longitude, and time for point data. This index will be created if the feature type has the geometry type Point and has a time attribute. This is used to efficiently execute queries with both spatial and temporal components.

2.4   Performance Comparison

In this section we provide some experimental results evaluating the performance of these NoSql databases[18]. With regard to the infrastructure used and since there is a need to work in a distributed mode, we have set up two clusters based on 4 virtual machine servers. Figure 1 illustrates our Elasticsearch cluster, a group of four nodes connected together.

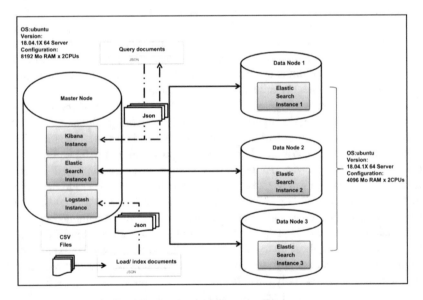

Fig. 1. Elasticsearch cluster architecture.

Each node has an Elasticsearch instance installed on it and may play one or more roles. Three data nodes dedicated for storing data and executing data-related operations such as search and aggregation. The fourth node has a master's role and is in charge of cluster-wide management and configuration actions such as adding and removing nodes.

---

[18] We have chosen to make this comparison study between Accumulo and Elasticsearch.

This node plays also an ingestion role, in the way that data arrives as a CSV files and is formatted in JSON using an open source data processing pipeline: Logstash[19]. In addition, we transform longitude/latitude to a geo_point object before injecting all the record into Elasticsearch data nodes.

Note that the index creation is based on the spatial hash function: GeoHash and is calculated during data insertion. Once done, data is ready to be queried using json on kibana.

For Accumulo cluster, we used four virtual machines with the same characteristics as Elasticsearch. According to Fig. 2, the cluster consists of a master node responsible for load balancing, failure management, table creation, table modification, and table deletion requests. And three Tabletservers nodes, each of them manages tablets partitions by receiving write instructions, persisting them to a write-ahead log and responding to reads from client queries. Data ingestion is done by the master node through Geomesa commands by loading a converter defined in a JSON-based instruction file that specifies instructions on how to turn longitude/latitude location information and convert them to the OGC standard Point type before storing them into Accumulo tablets.

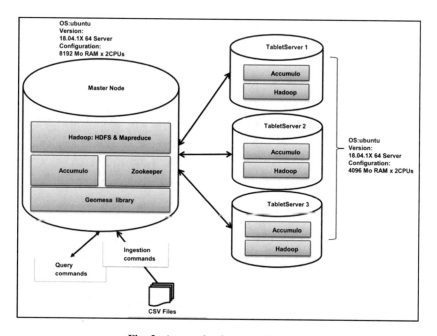

Fig. 2 Accumulo cluster architecture.

---

[19] https://www.elastic.co/products/logstash.

Data used in our experimentations are open data[20] from Telecoms and were shared by Zhang [18]. This data allows locating Telecom customers on the basis of their activity notably calls and SMS. For privacy concerns, all sensitive data was removed and all identifiable IDs have been replaced by serial numbers. The dataset is structured as the following structure: ID, time, latitude, longitude (Example: 0033336100, 10:17:52, 123.142705, 33.57902). We have reorganized this dataset by cutting it into several files of different volumes ranging from 1000 records to 1 billion records. These files are then loaded on Elasticsearch and Accumulo, respectively.

Figure 3 represents ingestion duration captured for each file. We can see that Accumulo performs better than Elasticsearch no matter the size of the ingested file. Indeed, Accumulo takes less than 8% of the time needed for Elasticsearch to handle 10 million records. In addition, Elasticsearch saturated repeatedly without being able to ingest the two files with 100 million and 1 billion records, while Accumulo ingested them without problems. Of course, ingestion durations are still high but regarding the current capacity of the machines used, they are acceptable. Note that we took care to have the same conditions between the two environments (no interruption, no other tasks in addition to the ingestion process).

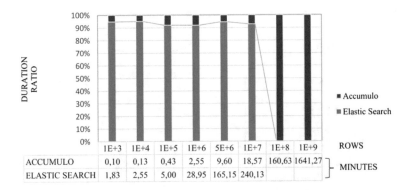

Fig. 3 Accumulo and elasticserach ingestion duration

The second comparison we have done is about spatial queries. We run a list of queries against spatial vector data previously loaded in both clusters, in order to collect geometry points satisfying to these conditions:

- Bounding Box Filter: Extracts point locations contained in a bounding box geometry with a given boundaries.
- Distance Filter: Extracts point locations that are at a given distance from a focal point of interest. We used a function that returns the 2D Cartesian distance between two geometries in the same units of the coordinate reference system (EPSG: 4236[21]) for both Accumulo and Elasticsearch.

---

[20] https://www-users.cs.umn.edu/~tianhe/BIGDATA/.

[21] https://spatialreference.org/ref/epsg/hu-tzu-shan/.

- Within Polygon Filter: Extracts point locations contained in a Polygon geometry. We used a constructor that converts a collection of geometry points to a polygon geometry and use it to filter data.
- KNN: Extracts for a focal point F, the K set of neighbours points closest to F based on a 2D Cartesian distance (EPSG: 4236)

Figure 4 shows the execution time in seconds taken by the four spatial queries already presented. We selected a 10 Million record dataset to do this exercise because Elasticsearch could not ingest more data. Results show that the execution time of these requests is very low on both NoSql databases. It took a few seconds despite the complexity of processing and the large volume of data. In addition, we notice that Accumulo is more efficient than Elasticsearch in executing all queries. For example, Accumulo needs only 20% of the time required by Elasticsearch to render knn and bounding box results.

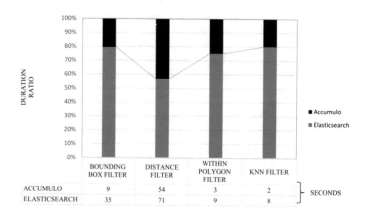

Fig. 4 Spatial queries execution durations

3 Conclusion

As mentioned earlier, management and storage of massive spatial data is challenged by the presence of big data constraints. We presented in this article a comparative study between three types of NoSql databases: document, key-value and column family. This case study allowed us to understand how we can take advantage of these new paradigms in the management and storage of spatial data. We also made an experimental comparison between Accumulo and Elasticsearch which allowed us to conclude that Accumulo is better than Elasticsearch on the data ingestion aspect. The same situation is observed when it comes to spatial queries: bounding box, distance, within polygon and KNN filters.

In our future work, we plan to deploy the architecture presented in this paper on more performant clusters in order to run ingestion processes and spatial queries on larger datasets. We will also include Redis and extend spatial filters presented in this paper with other spatial queries by including the time dimension.

References

1. Codd, E.F.: A relational model of data for large shared data banks. Commun. ACM **13**(6), 377–387 (1970)
2. Bachman, C.W.: Integrated data store. DPMA Q. **1**(2), 10–30 (1965)
3. Chamberlin, D.D., Boyce, R.F.: SEQUEL: a structured English query language. In: Proceedings of the 1974 ACM SIGFIDET (now SIGMOD) Workshop on Data Description, Access and Control, pp. 249–264. ACM (1974)
4. De Mauro, A., Greco, M., Grimaldi, M.: A formal definition of Big Data based on its essential features. Libr. Rev. **65**(3), 122–135 (2016)
5. Amirian, P., Basiri, A., Winstanley, A.: Evaluation of data management systems for geospatial big data. In: International Conference on Computational Science and Its Applications, pp. 678–690. Springer, Cham (2014)
6. DATA, semi-structured. SQL and NoSQL database comparison. In: Proceedings of the 2018 Future of Information and Communication Conference (FICC). Advances in Information and Communication Networks, pp. 294. Springer (2018)
7. Mukherjee, S.: The battle between NoSQL databases and RDBMS. University of the Cumberlands Chicago, United States (2019)
8. http://www.strozzi.it/cgi-bin/csa/tw7/i/en_us/nosql/homepage
9. Moniruzzaman, A.B.M., Hossain, S.A.: NoSQL database: new era of databases for big data analytics-classification, characteristics and comparison. arXiv preprint arXiv:1307.0191 (2013)
10. http://nosql-database.org/
11. Cormen, T.H., Leiserson, C.E., Rivest, R.L., et al.: Chapter 11: Hash tables. In: Introduction to Algorithms, pp. 221–252. Mit press and McGraw-Hill (2001). ISBN 978-0-262-53196-2
12. Procopiuc, O., Agarwal, P.K., Arge, L., et al.: Bkd-Tree: a dynamic scalable kd-Tree. In: International Symposium on Spatial and Temporal Databases, pp. 46–65. Springer, Berlin (2003)
13. Mpinda, S.A.T., Maschietto, L.G., Bungama, P.A.: From relational database to columnoriented NoSQL database: migration process. Int. J. Eng. Res. Technol. (IJERT) **4**, 399–403 (2015)
14. Chang, F., Dean, J., Ghemawat, S., et al.: Bigtable: a distributed storage system for structured data. ACM Trans. Comput. Syst. (TOCS) **26**(2), 4 (2008)
15. Böxhm, C., Klump, G., Kriegel, H.-P.: XZ-Ordering: a space-filling curve for objects with spatial extension. In: International Symposium on Spatial Databases, pp. 75–90. Springer, Berlin (1999)
16. Laksono, D.: Testing spatial data deliverance in SQL and NoSQL database using NodeJS fullstack web app. In: 2018 4th International Conference on Science and Technology (ICST), pp. 1–5. IEEE (2018)
17. Baralis, E., Dalla Valle, A., Garza, P., et al.: SQL versus NoSQL databases for geospatial applications. In: 2017 IEEE International Conference on Big Data (Big Data), pp. 3388–3397. IEEE (2017)
18. Zhang, D., Zhao, J., Zhang, F., et al.: UrbanCPS: a cyber-physical system based on multi-source big infrastructure data for heterogeneous model integration. In: Proceedings of the ACM/IEEE Sixth International Conference on Cyber-Physical Systems, pp. 238–247. ACM (2015)

Big Data for Internet of Things: A Survey on IoT Frameworks and Platforms

Amine Atmani[1,2]([⊠]), Ibtissame Kandrouch[1,2], Nabil Hmina[1,2], and Habiba Chaoui[1,2]

[1] Ibn Tofail University, Kenitra, Morocco
amine.atmn@gmail.com, mejhed90@gmail.com,
Ibtissame.Kandrouch@uit.ac.ma,
hmina@univ-ibntofail.ac.ma
[2] ADSI Team, System Engineering Laboratory, Kenitra, Morocco

Abstract. Ordinary objects that we use in our daily life have become now connected to the Internet and are getting even smarter. Wearable devices, thermostats, cars, door locks, lights, and more appliances are now connected over the Internet of Things. Therefore, the number of these smart things increases remarkably. At the present time, we talk about billions of connected devices and it is expected that this rapid growth will carry forward in an exponential way. This development has led to the investment in IoT applications that allow users interacting with all their devices, monitoring and controlling them remotely. Furthermore, the massive amount of data generated by connected devices and sensors should be transformed into actionable insights and predictions thanks to Big Data technologies for better user experience automation. In the interest of developing and implementing IoT applications, many Internet of Things frameworks and platforms are now designed. This paper provides a review of several available IoT frameworks and platforms. For each one of them, we discuss its architecture and its important features. Moreover, these frameworks are compared to each other depending on several criteria, such as: Security, data analytics, and support of visualization.

Keywords: Internet of Things · IoT · Big Data · Framework · Platform · Security

1 Introduction

Nowadays, Internet of Things is one of the fastest growing industries in the world. It is more and more affecting our daily life and developed and deployed almost in every field. The main application domains of IoT are: industry [1], healthcare [2], smart cities, e-commerce, environment [3], military [4] and energy [5]. Hence, this new connected world in full emergence forces companies to prepare and exploit new opportunities, and imagine new business models around the data generated by smart devices. This will allow to exploit these data in order to make the user's life easier.

We can define the Internet of Things (IoT) as a giant network, in which physical devices (things) are inter-connected and they can be controlled or monitored remotely.

M. Ezziyyani (Ed.): AI2SD 2019, LNNS 92, pp. 59–67, 2020.
https://doi.org/10.1007/978-3-030-33103-0_7

These devices are connected to the Internet, and they have the ability to interact with each other without human intervention. Therefore, they are considered to be smart and not only simple objects. There are various types of IoT devices (wearable devices, smart thermostats, IP cameras, robots, health monitoring devices etc.), and most of them are embedded with sensors, thus they can automatically detect events or changes in the surrounding environment, and transmit these data over network protocols. Big amount of data is collected from different IoT connected devices, then forwarded to servers (cloud) for analysis purpose, and finally results are shared with other devices in order to improve users experience.

Actually, there are multiple frameworks and platforms in the IoT market, which make easy the development and the deployment of IoT applications. However, it is a real challenge for developers to make the best choice of a framework, and this depends on their own visions.

Few studies have been covering IoT frameworks and platforms topic. Derhamy et al. [6] give a comparative analysis of several commercial IoT frameworks, based on the architecture approach, organizations support, supported communication protocols, hardware requirement, interoperability, security, governance, and support for rapid application development. Ammar et al. [7] give a review of several IoT frameworks, which includes the architecture, the specifications of smart applications and hardware, and the security features of each framework.

In this paper we present a survey of five well-known IoT frameworks, such as: AWS IoT, Azure IoT Suite, Iotivity, SmartThings, and Eclipse Kura, and we focus on Big Data and security features of each one of them.

This paper is structured as follow: In Sect. 2 we present the general IoT architecture. Then we discuss in Sect. 3 the relation between Big Data and IoT with the focus on Big Data processes that are employed in the IoT domain. Thereafter we present in Sect. 4 an overview of several available IoT frameworks and platforms. For each one we discuss its architecture and its important features. Section 5 gives a comparative analysis of theses frameworks depending on multiple criteria, such as: The framework type, pricing model, messaging protocols, support of virtual devices, data storage, data cleaning, data analytics, real-time data analytics, Machine Learning, data visualization and security features. Finally, we conclude the paper with Sect. 6.

2   IoT Architecture

There is no standard architecture for Internet of Things yet, but many researchers commonly present this architecture as three basic layers: Perception/Physical layer, Network layer and Application layer as shown in Fig. 1.

Perception Layer: It includes the hardware used in the IoT ecosystem (sensors and actuators). In this layer we can find several technologies such as Wireless sensor network (WSN), Radio-frequency identification (RFID) and Near-field communication (NFC). The major requirements for this layer are the support of the heterogeneity of IoT devices and the energy efficiency, thus sensors should be operational for gathering and transmitting data in real-time [8].

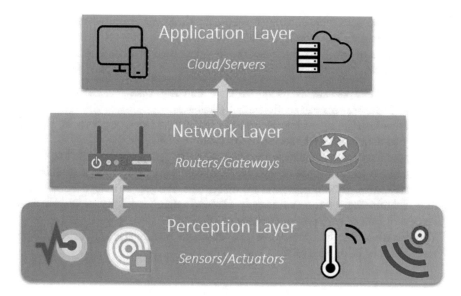

Fig. 1. IoT three-layer architecture

Network Layer: The second layer is responsible to ensure the communication of IoT devices with each other, in the case of Wireless sensor networks (WSNs) for example, or in some other cases, the communication directly with the cloud via a gateway. Data are collected by the perception layer and transmitted for analysis and decision-making purpose by several communication protocols, such as Bluetooth Low Energy (BLE), IEEE 802.15.4 standard and ZigBee in the case of low-power and low-bandwidth needs, also WIFI, 4G and 5G are used in this layer.

Application Layer: The third layer is the software part of the IoT architecture. It is responsible of presenting data after being collected and analyzed for a specific IoT application domain (e.g., healthcare, transportation, smart grid etc.). It is considered as the front end of the IoT architecture [9]. Several protocols are deployed in this layer, such as Constrained application protocol (CoAP), Message Queuing Telemetry Transport (MQTT), Hypertext Transfer Protocol (HTTP) etc.

3 Big Data and IoT

Nowadays, the number of connected devices increases exponentially. Thus, the amount of generated data from different sources in the smart environment is massive, hence the use of Big Data technologies in the IoT ecosystem is absolutely necessary. They are used to carry out a set of processes on generated data for applying actions on IoT devices, especially predictions, which will finally improve user's daily life.

Big Data classification according to its 5Vs model is suitable for the IoT case, because the size of generated data is immense (volume), and these data are coming from different sources with different types (variety), collected in real time (velocity), not certain (veracity) and worthy (value).

Four aspects of Big Data processes are the most used in the IoT domain such as: Storage, Data cleaning/cleansing, Analysis/Analytics and Visualization [10].

- **Storage:** Cloud storage is widely used in this Big Data feature, with favoring NoSQL Databases instead of Relational Databases.
- **Data cleaning:** This is an important phase that comes before analysis, in which the quality of data is verified and improved [11].
- **Analysis:** Clustering is the most used algorithm for data analysis in the IoT domain. In this phase various types of data are examined in order to extract valuable conclusions from it, which present a lot of benefits in every IoT application domain (e.g., minimizing traffic congestion, predicting future electricity needs, predicting epidemics, understand costumer needs etc.) [12].
- **Visualization:** In this phase, data is presented and interpreted using different methods after being analyzed.

4 IoT Frameworks

IoT frameworks helps developers for implementing and deploying IoT-based applications in the shortest time possible. It provides a collection of guiding rules, protocols and standards. For this survey we have chosen five frameworks to review.

AWS IoT: *Amazon Web Service IoT* is a platform which allows connected devices to communicate securely with each other and with the cloud. It offers several services like *AWS IoT Analytics*, which is capable of analyzing petabytes of generated data [13].

AWS IoT architecture has four principal components: *Device Gateway*, *Rules Engine*, *Device Shadows* and *Registry* [14]. The *Device Gateway* enables devices to interact with the cloud and with each other over MQTT protocol, with supporting the publish/subscribe programming model. The *Rules Engine* manages data delivery from publisher to subscriber devices depending on rules. The *Device Shadow* stores the last state of each device before being offline. It allows applications to interact with devices even when they are not online. The last component of *AWS IoT* architecture is the *Registry*. It is responsible of identifying connected devices by giving them a unique Id number.

Regarding the security of *AWS IoT* framework, the *Message Broker* gives each connected device mutual authentication, using different techniques, mainly the X.509 certificates. Also, authorization and access control process are guaranteed by the *Rules Engine*. Moreover, developers can create their own rules, using the *AWS Management Console Service*. Finally, all data traffic is encrypted over the Transport Layer Security (TLS).

Iotivity: *Iotivity* framework is an open source project, intended to Smart Home field. It facilitates Device-to-Device communication by using CoAP as messaging protocols, and provides a CoAP-HTTP proxy for allowing CoAP clients to interact with resources from HTTP servers [15]. Furthermore, it supports the publish/subscribe programming model [16].

Iotivity architecture is composed of the *Base Layer*, the *Service Layer* and the *Cloud Interface* [17]. The *Base Layer* includes an important feature (Discovery), which allow a device to find other nearby devices. Security is also implemented in this layer (authentication, securing data flow by using DTLS/TLS protocols and an access control policy, ensured by the *Security Resource Manager*), the last component of the *Base Layer* is *Messaging*. It allows the connectivity of IoT devices. Among the *Service Layer* functionalities there is an important one called *simulator*, which allows developers or users testing a set of devices before buying them.

Azure IoT Suite: It is an IoT cloud-based platform, released by Microsoft. It allows users to connect hundreds of heterogeneous devices with the cloud. It provides many functionalities, such as: The possibility of testing a set of virtual devices before buying them, Machine Learning and real-time data analytics *by Azure Stream Analytics*; this service allows developers to write their stream processing logic using SQL like language, with an embedded support for temporal logic, also to build real-time dashboards [18].

The backbone of *Azure IoT Suite* architecture is the *Azure IoT Hub*. It is the gateway between the cloud and the connected devices, and it allows them to communicate over multiple protocols, such as: HTTP, AMPQ or MQTT. It supports also other protocols that are adapted by the *Azure IoT Protocol Gateway*. In the *Azure IoT Suite* architecture there are two categories of devices: IP capable devices, that are linked directly to the *Azure IoT Hub,* because of implementing one of the three supported protocols, and PAN (Personal Area Network) devices, that are connected to the *Azure IoT hub* via the *Field Gateway,* since they are constrained. In the cloud side of the architecture we find two layers: *IoT Solution Backend*, which provide multiple cloud services and *Presentation and Business Connectivity*, which is responsible of data visualization.

As for the security of this framework, it is applied at three levels: The device, the connection and the cloud [19]. Two methods are used for the authentication of devices to the *Azure IoT Hub*: The first one is a security token method, based on a symmetric key, and the second one depends on X.509 certificates and a private key. For securing the data flow, TLS protocol is used to encrypt all communications. Lastly, access control policies for each security key are implemented at the *Azure IoT Hub* level.

SmartThings: It is an IoT platform released by Samsung. It is designed for Smart Home applications, and supports most of home automation devices. It allows users to connect, control and monitor their smart devices via a single mobile application. For developing and implementing *SmartThings* applications, Groovy is used as programming language [20].

SmartThing Hub is the backbone of the platform's architecture [21]. It is the intermediary between the smart devices and the cloud services that are accessible via a mobile application. It is compatible with multiple standards, such as: LAN, ZigBee and Z-Wave. *SmartThings Cloud Backend* provides several functionalities, such as: *Connectivity Management, SmartApp Management and Execution, Web UI* and *IDE.*

Security features of *SmartThings* framework are the following: OAuth/OAuth2 protocol is used for the authentication of smart devices. For access control, *Smart-Things Capability Model* is provided for limiting smart applications assessing all devices capabilities. Moreover, *Kohsuke sandbox* technique is applied for denying untrusted Groovy scripts. Lastly SSL/TLS protocol is provided for encrypting communications [7].

Eclipse Kura: It is an open source IoT framework based on Java/OSGI. It is dedicated for building IoT gateways that contain Machine-to-Machine (M2M) applications. It offers a wide range of APIs for developers, with an abstraction and isolation from the complexity of the hardware. Furthermore, this framework supports only Linux-based devices.

The components of *Eclipse Kura* architecture are the following: Firstly, the *Device Abstraction*, which gives developers the access for many devices by OSGI services, with abstracting the hardware. Then *Basic Gateway Services,* which includes multiple services (watchdog, clock, embedded database etc.). Also, *Network Configuration,* which allows the configuration of interfaces available in the gateway (Ethernet, Wi-Fi or Cellular). *Connectivity and Delivery,* used to make easy the development of applications that are connected with a remote cloud over MQTT protocol. *Remote Management,* responsible of the deployment, upgrade, configuration and management of the IoT applications. *Application,* which offers a data flow programming tool for data storage and aggregation. Finally, *Administration GUI,* which is a web interface dedicated for managing the gateway [22].

Concerning the security of this framework, ESF is an open source tool added to *Kura* framework in order to enforce its security; it provides runtime policies for limiting the execution of some services. Moreover, SSL protocol is implemented for encrypting all communication and for a mutual authentication of IoT devices, using a couple of certificates (public and private key) [23].

The following table gives a comparison of the five underlying IoT frameworks and platforms (Table 1).

5   Discussion

We can classify the five underlying frameworks in two categories: open source frameworks (Iotivity, SmartThings and Eclipse Kura) and paid solutions (AWS IoT and Azure IoT Suite).

The two latter frameworks offer a wide range of services; all Big Data features are included (Data storing, Real-time data analysis, data cleaning, data visualization using dashboards), with the support of Machine Learning and the storage of a big amount of data in No SQL databases. Also, all security measures are applied at different layers of their architecture. Furthermore, they all support MQTT which is a lightweight protocol in terms of small code footprint and low bandwidth requirement. However, these two frameworks need costs for each used service.

Concerning the open source frameworks, they provide several features for building IoT applications easily: All the three underlying frameworks implement different security features either for devices authentication (with OAuth or SSL protocol) or for access control policies, or for data flow encryption. They also support CoAP, which is an important protocol of IoT, designed for constrained devices, in addition of MQTT protocol (Eclipse Kura case). There is a lack in terms of some Big Data features, such as data analysis and data visualization. Otherwise, it is possible to combine these frameworks with other stream analytics frameworks and cloud platforms.

Table 1. Comparison of different IoT frameworks and platforms

Framework	AWS IoT	Iotivity	Azure IoT Suite	SmartThings	Eclipse Kura
Open source	No	Yes	No	No	yes
Pricing model	Pay-as-you-go	–	Pay-as-you-go	–	–
Messaging protocols	HTTP and MQTT	CoAP and HTTP	HTTP, MQTT and AMQT	HTTP	MQTT and CoAP
Virtual devices	Yes	Yes	Yes	Yes	No
Storage	Relational and No SQL databases (Amazon DynamoDB, Amazon S3, Amazon RDS)	Yes	Relational and No SQL databases (Azure Cosmos DB, Azure SQL) Database)	Yes	Yes
Data cleaning	Yes	–	Yes	No	Yes
Data analytics	Yes (AWS IoT Analytics)	Yes	Yes (Azure Stream Analytics)	No	No
Real-time data analytics	Yes	No	Yes	No	No
Machine learning	Yes	No	Yes	No	No
Data visualization	Yes (AWS IoT Dashboard)	No	Yes	No	No
Authentication	X.509 Certificates + AWS IAM + AWS Cognito	OAuth2	X.509 Certificates + private key	OAuth/OAuth2	SSL (public and private key)
Access control	Rules Engine	Security Resource Manager	Access control policies for Azure IoT Hub	SmartThings capability model + Kohsuke sandbox technique	Runtime policies by ESF
Communication security	TLS	DTLS/TLS	TLS	SSL/TLS	SSL

6   Conclusion

The development of Internet of Things world has led to the creation of sophisticated IoT applications which include all users requirements such as: Security, privacy, data analytics for the extraction of important insights from data etc. This is the role of IoT frameworks and platforms which provides a set of guiding rules and ready-to-use features that accelerate the process of developing and deploying IoT applications.

In this survey, we have brought into focus the key features of five different IoT frameworks and platforms from open source and paid solutions. Then we have presented a comparative analysis of theses frameworks depending on multiple criteria, with focusing on the use of Big Data features and the security of each one of them, because of their necessity in IoT applications. Therefore, this study can give developers a clear vision of which framework is the most suitable for their future projects.

References

1. Da Xu, L., He, W., Li, S.: Internet of things in industries: a survey. IEEE Trans Ind. Inform. **10**, 2233–2243 (2014). https://doi.org/10.1109/TII.2014.2300753
2. Yin, Y., Zeng, Y., Chen, X., Fan, Y.: The internet of things in healthcare: an overview. J. Ind. Inf. Integr. **1**, 3–13 (2016). https://doi.org/10.1016/j.jii.2016.03.004
3. Asghari, P., Rahmani, A.M., Javadi, H.H.S.: Internet of things applications: a systematic review. Comput. Netw. **148**, 241–261 (2019). https://doi.org/10.1016/j.comnet.2018.12.008
4. Raglin, A., Metu, S., Russell, S., Budulas, P.: Implementing internet of things in a military command and control environment. In: Next-Generation Analyst V, vol. 10207, p. 1020708 (2017). https://doi.org/10.1117/12.2265030
5. Wong, S., Pinard, J.P.: Opportunities for smart electric thermal storage on electric grids with renewable energy. IEEE Trans. Smart Grid **8**, 1014–1022 (2017). https://doi.org/10.1109/TSG.2016.2526636
6. Derhamy, H., Eliasson, J., Delsing, J., Priller, P.: (2015) A survey of commercial frameworks for the internet of things. In: Conference on Emerging Technologies Factory Automation (ETFA), October 2015. https://doi.org/10.1109/etfa.2015.7301661
7. Ammar, M., Russello, G., Crispo, B.: Internet of things: a survey on the security of IoT frameworks. J. Inf. Secur. Appl. **38**, 8–27 (2018). https://doi.org/10.1016/j.jisa.2017.11.002
8. Li, S., Da Xu, L., Zhao, S.: 5G internet of things: a survey. J. Ind. Inf. Integr. **10**, 1–9 (2018). https://doi.org/10.1016/j.jii.2018.01.005
9. Romdhani, I.: Architecting the Internet of Things (2011). https://doi.org/10.1007/978-3-642-19157-2
10. Ge, M., Bangui, H., Buhnova, B.: Big data for internet of things: a survey. Futur. Gener. Comput. Syst. **87**, 601–614 (2018). https://doi.org/10.1016/j.future.2018.04.053
11. Karkouch, A., Mousannif, H., Al Moatassime, H., Noel, T.: Data quality in internet of things: a state-of-the-art survey. J. Netw. Comput. Appl. **73**, 57–81 (2016). https://doi.org/10.1016/j.jnca.2016.08.002
12. Hashem, I.A.T., Yaqoob, I., Khan, I., Vasilakos, A.V., Imran, M., Ahmed, A.I.A., Ahmed, E.: The role of big data analytics in internet of things. Comput. Netw. **129**, 459–471 (2017). https://doi.org/10.1016/j.comnet.2017.06.013
13. AWS IoT: AWS IOT services (2019). https://aws.amazon.com/iot/. Accessed 1 Mar 2019
14. AWS IoT: Components of AWS IoT architecture (2019). https://aws.amazon.com/iot-core/features/. Accessed 1 Mar 2019
15. Iotivity: COAP-HTTP proxy (2019). https://wiki.iotivity.org/coap-http_proxy. Accessed 1 Mar 2019
16. Iotivity: Publish/Subscribe model (2019). https://wiki.iotivity.org/message_queue_mq_for_publish-subscribe_interactions. Accessed 1 Mar 2019
17. Iotivity: Iotivity Architecture (2019). https://wiki.iotivity.org/architecture. Accessed 1 Mar 2019
18. Azure IoT Suite: Azure Stream Analytics (2019). https://azure.microsoft.com/en-us/services/stream-analytics/. Accessed 1 Mar 2019
19. Azure IoT Suite: Azure IoT Suite Security (2019). https://docs.microsoft.com/en-us/azure/iot-fundamentals/iot-security-deployment. Accessed 1 Mar 2019

20. SmartThings: Groovy programming language (2019). https://docs.smartthings.com/en/latest/getting-started/groovy-for-smartthings.html. Accessed 1 Mar 2019
21. SmartThings: SmartThings architecture (2019). https://docs.smartthings.com/en/latest/architecture/index.html. Accessed 1 Mar 2019
22. Eclipse Kura: Eclipse Kura architecture (2019). http://eclipse.github.io/kura/intro/intro.html. Accessed 1 Mar 2019
23. Eclipse Kura: Eclipse Kura security (2019). http://eclipse.github.io/kura/config/ssl-configuration.html. Accessed 1 Mar 2019

Current State and Challenges of Big Data

Yassine Benlachmi[(⊠)] and Moulay Lahcen Hsnaoui[(⊠)]

Higher School of Technology, ISIC ESTM, LMMI Laboratory,
ENSAM Moulay-Ismail University, Meknes, Morocco
yassin040@gmail.com, myhasnaoui@gmail.com

Abstract. Big Data is the complex, bulky, growing set of data coming from independent sources. In today's modern age Big data has an essential part in nearly every field of human life including science, engineering, social, biological and biomedical departments. In the following paper importance of big data, stream learning, deep learning, Hadoop and its application are discussed.

Keywords: Big Data · Machine learning · Stream learning · Hadoop

1 Introduction

Big Data which by name tells that it will be huge in volume is the technology of today as well as future. It's ruling the world now and its going to rule the world in near and in far future too as per scientists and researchers. So, what is Big Data? The data which cannot be stored and processed on traditional systems like relational database management systems is known as Big Data. The size of data generated by humans, machines and their interactions on social media itself is huge. Researchers predicted that 40 Zettabytes (40,000 Exabytes) will be generated by 2020, which is an increase of 300 times from 2005. Let me tell you few fun facts. In 5 years, there will be more than 50 Billion smart connected devices in the world. We are creating 2.5 Quintillion Bytes of Data Everyday. There will be 6.1 Billion global smartphone users. Data generated every minute is huge in the number itself. Would you believe me If I said Facebook users like 4,166,667 posts per minute, Twitter users send 347,222 tweets per minute, Reddit users cast 18327 votes per minute, Instagram users like 1,736,111 posts per minute, YouTube users upload 300 h of new video per minute.

Big data can come from any source and its growth never ends apparently. It occupies tons of space from gigabytes to exabytes or even zeta bytes these days. Due to its huge size, big data is not stored through conventional ways and in conventional databases. In Table 1 below the difference between Traditional and Big data is shown.

The remainder of this paper is organized as follows. Section 1 presents different types and sources of big data that exists today. It also discusses applications of big data in different domains like testing, social systems, machine learning, deep learning, etc. Section 2 talks about what are the problems in storing and processing Big Data. Section 3 introduces solutions to big data challenges. Section 4 introduces how big data is leveraged by streaming and deep learning models in Artificial Intelligence. Section 4 talks about Hadoop, which is the most popular framework to store and process big data. Finally, we terminate the paper with a conclusion and perspectives.

© Springer Nature Switzerland AG 2020
M. Ezziyyani (Ed.): AI2SD 2019, LNNS 92, pp. 68–80, 2020.
https://doi.org/10.1007/978-3-030-33103-0_8

Table 1. Difference between traditional and big data

Components	Big data	Traditional data
Queries	Largely abandoned SQL	Traditional SQL
Architecture	Distributed	Centralized
Data types	Structured, semi- structured and unstructured	Structured
Data model	No schema	Fixed schema
Data relationship	Unknown or complex	Known relationship
Data volume	Petabytes or Exabytes	Terabytes
Data traffic	More	Less
Data integrity	Less	High

2   Source of Big Data

Learning through observation is the basic and main instinct of mankind. From the starting, we have tried to rule every aspect of life by learning through observation and mistakes. To learn from those observations and mistakes we need the data to understand the behavior and happenings of the incident we are trying to avoid or to increase its frequency. For this kind of learning most of the time data is not in small size, it's more than huge that's why we call it big data. So, the first step is always data acquisition. Some data sources such as sensors and other electric circuits produce an enormous amount of raw data, but the problem is that most of that data is not beneficial for us and it can be filtered out. So, the first step in the following dimension designing and applying suitable filters to get the required useful data. The second step is information extraction from the raw data after applying the suitable filter. In the third step, the real challenge begins which is analyzing the filtered-out information and it is much more challenging than just identifying, locating and citing data [1].

2.1   The Need of Big Data?

In simple words, Big Data plays a most important role in observing behaviors and patterns of any system over a long period of time and it helps us to foresee the future of the system but the way of using big data could vary. The applications of Big Data are in every field of life but few of them are discussed below.

2.2   Software Testing

Software testing is all about observation of the behavioral pattern of the software that how it is going to behave in different situations and under variable conditions so that we could find out bugs and errors present in there and we could improve and optimize the software. To get the required output from any system first we have to put the input and in this case, it's always data and not just a small amount of that but in a huge amount of data and information about every feature and option present in the software that how it's working in variable scenarios. Just like we have discussed before that in sensors and electric circuits most of the raw data is not useful same thing goes here, every data produced by software is not useful we have to filter out the required data and then we have to study it in order to get an efficient optimized app. Figure 1 below shows the process of software testing.

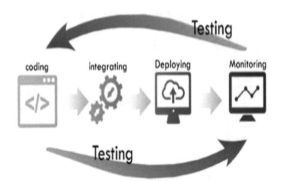

Fig. 1. Process of software testing

2.3 Social Systems and Social Attitude of Society

As the technology is advancing day by day the nature of mankind is also changing but the challenge here is that this transformation of attitude is not very rapid and it's not easy to study as well. For example, a study is required to how people used to react to the idea of feminism in the 20th century vs. how they react now so that we could able to understand the level of change during past years. For this purpose, we need the opinion of people and there are billions of people present in the world and here Big Data comes into action. Just like this in the field of medical sciences, the disease patterns are studied with the help of Big Data to reduce and eliminate the causes of disease in particular region of the world. Figure 2 given below shows a similar study held in Sweden for prostate cancer.

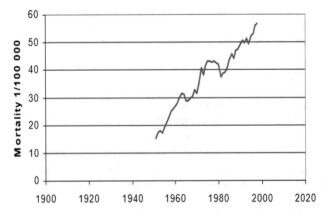

Fig. 2. Development of prostate cancer death rates in Sweden since 1951.

2.4 Big Data and Machine Learning

The main goal in today's age is to achieve command over Artificial Intelligence (AI) and Machine learning is the core component of AI. Machine learning is basically providing the ability of self-improving and learning to the system without being explicitly programmed.

In Machine Learning (MI) the main input that is required by the system to improve itself is the continuous supply of data. In MI the system automatically filters the data and extracts the required information then analyzes it. From that extracted information it observers and takes steps to automatically improve itself. Machine learning helps us to overcome the challenges that how a workflow can be automated and modernized [2]. There is no rules-based programming in Machine learning and system creates algorithms for itself by analyzing the data and recognizing the patterns in it. Machine Learning and Big Data are related to each other as they both are the part of data sciences and the common goal of both is to learn through data by analyzing it. As it is mentioned above Big Data is used to identify the patterns in Data and the trends of people from which data is coming whereas Machine Learning uses all this to improve the system automatically. It can also be said that Big Data analysis is the preprocessing step of Machine Learning.

Of late, there has been a lot of focus on automation and machine learning. Machine learning is used on massive data (big data) to understand the correlations and relationships between disparate data sources. It is ideal for exploiting the opportunities hidden [21] in big data. And the best part is, unlike traditional systems, Machine learning is best suited when you have a huge amount of data, and that's why Machine learning and Big Data [23] gel together easily. The more the data is, the better model will be created using Machine Learning. The machine learning models which are deployed in production these days are re-evaluated after a certain frequency by their creators. In case of big data production environments, when the architecture grows with time, machine learning models perform better with higher accuracy as more data is getting collected. Hence, to have collaboration between Big Data and Machine Learning is highly recommended.

Artificial Intelligence (AI) is based upon Machine Learning as it always tries to improve itself by learning through the user's behavior towards it. Analysis of data and identifying the user's patterns never stops for AI apparently as it constantly tries to get better and better. The example of the most commonly used kind of AI are software based mobile phone's assistants i.e. Apple's Siri[1], Microsoft Cortana[2] and Google Assistant[3]. This mobile assistant software uses machine learning to analyze the user's behavior, his daily routines, his type of works and reshape itself to needs of the user. They also constantly send data back to their companies where it's analyzed to improve the overall software composition. Corporate sector is using MI to enhance their systems

---

[1] https://www.apple.com/siri/.

[2] https://www.microsoft.com/en-us/cortana.

[3] https://assistant.google.com/.

and heavy amount of research papers are being published on AI and MI. Figure 3 below shows the rapid increase of research papers published on AI and MI.

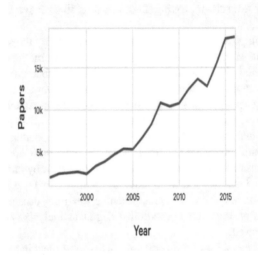

Fig. 3. Annually published AI Papers from 2000 to 2015

Figure 3 shows the upsurge in investment in AI startups based on Machine learning from 1995 to 2015 that shows how much corporate sector is showing interest in AI and MI. Figure 4 shows the upsurge in investment in AI startups based on Machine learning from 1995 to 2015 that shows how much corporate sector is showing interest in AI and MI.

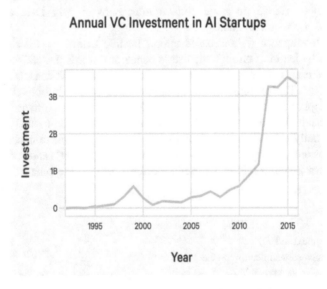

Fig. 4. Annual VC investment in AI startups

The next chapter in this paper talks about the several challenges with Big Data such as storage, security, processing etc.

3   Challenges of Big Data

Issues in the Big Data we faced it include the issues of the management, Security, Processing [4] and Storage [3–5] and they are briefly discussed below.

3.1   Management Issues

Big Data management [6] is the collection of large amounts of Structured, unstructured and semi-structured data from any organization or system. The main target of Big data management is to guarantee standardization, quality, data ownership, documentation and approachability of the data set [3]. The quality of the data any organization working on is very important to find meaningful insights out of it. The 4th V of Big data is called Veracity [19], which means accuracy. In Big data, the data coming in is un-structured and sometimes it's inaccurate or junk data. To work on inaccurate data to find meaningful insight is impossible. Due to the size of data, management is not an easy task and it requires a lot of workforce to handle it.

3.2   Storage Issues

Data generated in the past 2 years is more than the previous human race history in total. By 2020, total digital data will grow to 44 Zettabytes or 44 trillion Gigabytes approximately. By 2020, more than 1.7 MB of new info will be created every second for every person. You can't store this much Big Data into your traditional RDBMS like Oracle, MySQL or Microsoft SQL Server. There needs to be another solution for storing Big data. Virtualization in Big data is used for the storage purpose. In order to get that some software tools like NoSQL, Apache Drill, Horton Works [7], Map reduce are used which use complex algorithms and structures to manage and sort data. This increases the cost, it requires a lot of time and workforce. Storage is never been cheap so storage of such an enormous amount is always an issue in every case of storing data.

3.3   Processing Issues

Even if the storage challenge of Big Data gets solved, processing it is a very big challenge as the data is present in different formats. It can be structured, semi-structured and unstructured. Now analyzing unstructured data using our traditional RDBMS is impossible, as they work only with structured data. Due to the size of the data a massive amount of time it requires processing data which is in Petabytes, Exabytes or sometimes in Zettabytes. It also requires a huge amount of processing power which is really expensive and must be maintained. The technology also changes with time rapidly so it must be up to date that again requires money in every few years.

3.4   Security Issues

Data is considered as an important and expensive asset of any organization or system. That's why it can get stolen or misused by unwanted entities that is why it must not be vulnerable to threats and it is really a tough issue to handle a data securely. Hackers can steal data by attacking the data storage by sending attacks. In the case of Big Data, since the data is huge in size, making is secure [22] is another challenge. It includes user/group access permissions, user/group authentications, encrypting the credentials, recording data history, etc. Denial of service [8], Spoofing attack and Brute force attack [9] are some of the most commonly used attacks by hackers and it can steal data and valuable processed information which could be misused afterward. For this purpose, keeping the data security must be the top priority, in order to enhance the security, Framework techniques and vigorous algorithms must be developed. There are many software tools that could help in increasing the security of the Big Data, these tools include Hadoop and NoSQL, and so on.

4   Solutions to Big Data Challenges

There are several Big Data frameworks and tools in the market which can handle the above challenges of Big Data.

Apache Hadoop is a framework that allows the distributed processing of large data sets across clusters of commodity computers using a simple programming model. It solved the storage issue. In Hadoop, Data storing is distributed and easily scalable. Big Data is split into blocks (128 MB each) and is distributed in nodes. It also solved the processing issue through MapReduce. In Hadoop, the processing time is reduced drastically. Processing (code) is sent to the data (blocks) and is executed locally on the nodes. For real-time use cases, Apache Spark is the technology that is being used by all the industry leaders to do real-time analytics. Spark Streaming is used to stream real-time data from various sources like Twitter and perform powerful analytics to help businesses. NoSQL databases like HBase [24], Cassandra, MongoDB can store all the unstructured data with ease, which relational database systems cannot. Tools like Kerberos offer security to big data clusters and architectures. Big Data and Artificial Intelligence leverage each other.

5   Relationship Between Big Data and Learning Path

5.1   Big Data and Stream Learning

Machine learning has been discussed above so now we will move towards Stream learning. Most of the data these days is produced in the shape of stream, stream processing takes the data in the form of continuous stream and processes it as it arrives. Example of stream processing includes sensors in industry, System/Software/ Components Log files and online streaming of data/information that requires real-time processing of massive data from the large volume. One of the most popular and abundantly used platforms for real-time processing is the storm which is an open source that means free for everyone to use. It is scalable and provides a very fault tolerant platform for streaming data without any limits which means if a fault or any problem appears it can handle by itself and the program will not crash.

Streaming model has been made by the researchers and in the following model, data is received in the form of stream at very high speed. When the data is received at the instant of time immediately the algorithms are applied to process the data and results are produced. For this rapid action, algorithms for streaming purposes use probabilistic data structures so that immediate approximate answers could be provided [10].

5.2 Big Data and Deep Learning

Deep Learning (DL) or it is normally called deep structured learning is a branch of Machine Learning (ML) that depends on a collection of algorithms that try to design high-level concepts in data, [11, 12]. Deep learning is also known as hierarchical learning. As Deep Learning is a branch of Machine Learning and Machine Learning requires Big Data to improve itself same case goes for Deep Learning. Deep Learning also requires gigantic amounts of data so that algorithms could be developed. In Deep Learning system has given tons of examples for a problem and it tries to develop an algorithm for it that helps to decide the solution of the relatively same problems by itself, automatically. The latest and successful example of deep learning is facing recognition system of Facebook; DeepFace. It used almost 400 million pictures of users for training of the system and reached the accuracy of almost 97.3%. The Pictures which were given as input in the following example was Big Data which the system analyzed through deep learning process and the patterns were identified through the analysis which helped the system to improve itself and achieve the said accuracy of 97.3%. The algorithms on which Deep learning depends develop a superimposed, ordered architecture of learning and characterizing data. This hierarchical learning design is based on the nature's super design which is in the form of human brain [13–15]. It follows the process of deep learning which is used in the primary sensory areas of neocortex (a part in the human brain) [11, 16]. Algorithms of deep learning plays a vital part in dealing with the huge volume of unsupervised data [17, 18].

6 Hadoop

HADOOP Fig. 5 is the abbreviation of Highly Archived Object-Oriented Programming. It is a freeware (Open Source) Java framework which is vital to store, access, modify and gain large resources from Big data in a distributed order at low cost and with high degree of fault tolerance which means it can handle most of the faults on its own, automatically. The persons who created Hadoop were Goug Cutting and Mike Cafarella in the year of 2005. Hadoop is capable of handling massive loads of data which could be from any organization or from different systems like Images, Videos, Files, Project Files, Folders, Sensor data, Hadoop can store all these data from any system. Components of Hadoop includes Sqoop, Chukwa, Avro, HBase, Hive, Flume, Oozie, Lucene, Pig, and Zookeeper [6].

Most of the time Hadoop had been misunderstood as a database, well it is not a database, it is a software ecosystem that allows for massively parallel computing. It allows data to be spread across thousands of servers without lowering the performance using NoSQL distributed databases like HBase.

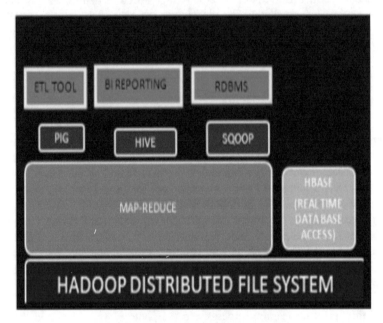

Fig. 5. Hadoop distributed file system

Some of the most prominent features of Hadoop are Documentation, location awareness, source code, and work scheduling, Hadoop 2 has two core components - HDFS and Mapreduce (YARN in Hadoop 2.x). HDFS stands for Hadoop Distributed File System which is used for storage and consists on namenode and datanode daemons. MapReduce is for processing in hadoop and consists of ResourceManager and Node Manager.

6.1 HDFS Architecture

As it is discussed before that Hadoop file system is a Java portable b which provides us with higher degree of scalability, prominent reliability and distribution in the Hadoop environment. The communication between nodes is done by procedure calls. In the case that the block is not working, or it is lost then a duplicate of the same block is created by name block. To identify the current state data node is always maintained by timestamp. Node can be fixed regularly in the case of any possible failure. More than 1000 nodes are allowed by HDFS for a single carrier signal.

HDFS runs on commodity hardware and assumes high failure rates of the components. It works well with lots of large files with hundreds of gigabytes and terabytes in size. It has been built around the idea of "write-once, read-many-times" because high throughput is more important than low latency. Hadoop Distributed Filesystem operates on top of an existing filesystem. Files are stored as blocks, the default block size is 64 MB in hadoop 1.x and 128 MB in hadoop-2.x. It provides reliability through replication, each block is replicated across several Data nodes. Namenode stores metadata and manages access. No data caching due to large datasets.

6.2 MapReduce

MapReduce is the core of Hadoop. It is a programming paradigm which gives scalability across thousands of servers in the Hadoop cluster. If you divide the term MapReduce, it will be Map ad Reduce; these are two separate tasks that are performed in Hadoop. The first task is the map job, at the level, the data taken will be converted to another data set, so that the individual elements are broken down into tuples (key/value pairs). The reduce job receive the output from a map as input and combines those data tuples into a smaller set of tuples. Firstly, map job is performed and then the reduce job. Sometimes only map job is enough to process the data in Hadoop.

MapReduce Fig. 6 is a method for distributing a task across multiple nodes. Each node processes data stored on that node to the extent possible.

The processing part of Hadoop ecosystem [24] is MapReduce, it is a computing model that processes a huge amount of data and spreads the computation across a huge number of servers in a Hadoop cluster. Two biggest Advantages of MapReduce are taking processing to the data and Processing data in parallel. When a MapReduce job starts the input data is distributed to nodes, ach map task works on a "split" of data, mapper outputs intermediate data. Data exchange between nodes in a "shuffle" process, intermediate data of the same key goes to the same reducer and finally, reducer output is stored.

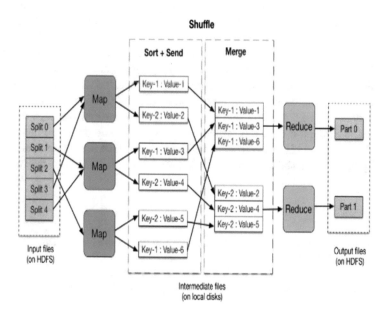

Fig. 6. Mapreduce process

A running Map Reduce job consists of various phases such as Map -> Sort -> Shuffle -> Reduce. The primary advantages of abstracting your jobs as MapReduce running over a distributed infrastructure like CPU and Storage are:

- Automatic parallelization [25] and distribution of data in blocks across a distributed, scale-out infrastructure.
- Fault-tolerance [26] against failure of storage, compute and network infrastructure
- Deployment, monitoring and security capability
- A clean abstraction for programmers

Its incredible processing technique can process a huge amount of data that might take only 3 min to process it when distributed across a large Hadoop cluster [28] of commodity servers in parallel whereas it will take 20 h of processing time on a centralized relational database system.

7   Conclusion

Big Data is needed to observer the behaviors and patterns of system over long duration of time. It is the core component of Machine Learning and AI. It is also playing a vital role in Stream learning and deep learning and its usefulness cannot be denied. Hadoop (Highly Archived Distributed Object-Oriented Programming) is an open source (free for everyone) Java framework which is used to store, access and gain massive resources from Big Data. Hadoop comprises of two main parts HDFS provides Storage of data and it also provides Map Reduce.

Most organizations are generating big data, but they were not aware of harnessing Big Data for their profits earlier. But with growing big data trend, 2019 and 2020 will be crucial years for Big Data analytics. Lot of operationalization of big data pipeline with take place across organizations leveraging Big Data. According to Forbes [20], by 2027 the Big Data market revenue will be around $103B. Also, as the industry is moving towards Artificial Intelligence, machine learning and deep learning models will get involved in Big Data implementations, creating better and smarter products across all domains – Retail, Manufacturing, Healthcare, Finance, Telecom, etc. Now that it is evident that Big Data domain will grow exponentially, I will work in identifying the right Big data tools [27] in my future work. There are several big data tools in the market but selecting the right tool which would be fit for any proposed architecture is difficult. Few popular big data tools on which I will continue my research are Hadoop, Spark, Sqoop, Flume, Kafka, Cassandra. do simulations and compare the results.

References

1. Labrinidis, A., Jagadish, H.: Challenges and opportunities with big data. Proc. VLDB Endowment 5, 2032–2033 (2012). https://doi.org/10.14778/2367502.2367572
2. Stouky, A., Jaoujane, B., Daoudi, R., Chaoui, H.: Improving Software Automation Testing Using Jenkins, and Machine Learning Under Big Data (2019). Accessed 24 Apr 2019

3. Saraladevi, B., Pazhaniraja, N., Paul, P., et al.: Big data and hadoop-a study in security perspective. Procedia Comput. Sci. **50**, 596–601 (2015). https://doi.org/10.1016/j.procs.2015.04.091

4. Ji, C.: Big data processing in cloud computing environments. In: International Symposium on Pervasive Systems, Algorithms, and Networks

5. Song, Y.-S.: Storing Big Data- The rise of the Storage Cloud (2012)

6. Philip, R.: Managing Big Data. TDWI research (2013)

7. Hortonworks: What is Apache Spark. In: Hortonworks (2019). https://hortonworks.com/apache/spark/. Accessed 24 Apr 2019

8. Shay, C.: Application Denial of Service. Hack tics Ltd (2007)

9. Daniel, J.: Understanding brute force. National Science Foundation, Chicago (2019)

10. Bifet, A., De Francisci, M.: Big data stream learning with SAMOA. In: IEEE International Conference on Data Mining Workshops, ICDMW 2015, pp. 1199–1202 (2015). https://doi.org/10.1109/ICDMW.2014.24

11. Hinton, G.E., Osindero, S., Teh, Y.W.: A fast learning algorithm for deep belief nets. Neural Comput. **18**(7), 1527–1554 (2006)

12. Bengio, Y.: Learning deep architectures for AI. Found. Trends® Mach. Learn. **2**(1), 1–127 (2009)

13. Bengio, Y., LeCun, Y.: Scaling learning algorithms towards AI. Large-Scale Kernel Mach. **34**(5), 1–41 (2007)

14. Bengio, Y., Courville, A., Vincent, P.: Representation learning: a review and new perspectives. IEEE Trans. Pattern Anal. Mach. Intell. **35**(8), 1798–1828 (2013)

15. Arel, I., Rose, D.C., Karnowski, T.P.: Deep machine learning-a new frontier in artificial intelligence research [research frontier]. IEEE Comput. Intell. Mag. **5**(4), 13–18 (2010)

16. Bengio, Y., Lamblin, P., Popovici, D., Larochelle, H.: Greedy layer-wise training of deep networks. Adv. Neural. Inf. Process. Syst. **19**, 153 (2007)

17. Hordri, N.F., Yuhaniz, S.S., Shamsuddin, S.M.: Deep Learning and Its Applications: A Review (2016)

18. Jan, B., Farman, H., Khan, M., Imran, M., Islam, I., Ahmad, A., Ali, S., Jeon, G.: Deep learning in big data analytics: a comparative study. Comput. Electr. Eng. (2017). https://doi.org/10.1016/j.compeleceng.2017.12.009

19. Demchenko, Y., Grosso, P., de Laat, C., Membrey, P.: Addressing big data issues in scientific data infrastructure. In: 2013 International Conference on Collaboration Technologies and Systems (CTS), San Diego, CA, pp. 48–55 (2015). https://doi.org/10.1109/cts.2013.6567203

20. Louis Columbus Contributor @Forbes, 23 May 2018. https://www.forbes.com/sites/louiscolumbus/2018/05/23/10-charts-that-will-change-your-perspective-of-big-datas-growth/#2c9ba09b2926

21. El-Gayar, O., Timsina, P.: Opportunities for business intelligence and big data analytics in evidence based medicine. In: 2014 47th Hawaii International Conference on System Sciences, Waikoloa, HI, pp. 749–757 (2014). https://doi.org/10.1109/hicss.2014.100

22. Wu, J., Ota, K., Dong, M., Li, J., Wang, H.: Big data analysis-based security situational awareness for smart grid. IEEE Trans. Big Data **4**(3), 408–417 (2018). https://doi.org/10.1109/tbdata.2016.2616146

23. Alfred, R.: The rise of machine learning for big data analytics. In: 2016 2nd International Conference on Science in Information Technology (ICSITech), Balikpapan, p. 1 (2016). https://doi.org/10.1109/icsitech.2016.7852593

24. Mazumder, S., Dhar, S.: Hadoop ecosystem as enterprise big data platform: perspectives and practices. Int. J. Inf. Technol. Manage. **17**(4), 334–348 (2018). Inderscience Enterprises Ltd

25. Vora, M.N.: Hadoop-HBase for large-scale data. In: Proceedings of 2011 International Conference on Computer Science and Network Technology, Harbin, pp. 601–605 (2011). https://doi.org/10.1109/iccsnt.2011.6182030
26. Mondal, K., Dutta, P.: Big data parallelism: challenges in different computational paradigms. In: Proceedings of the 2015 Third International Conference on Computer, Communication, Control and Information Technology (C3IT), Hooghly, pp. 1–5 (2015). https://doi.org/10.1109/c3it.2015.7060186
27. Kasu, P., Kim, Y., Park, S., Atchley, S., Vallée, G.R.: Design and analysis of fault tolerance mechanisms for big data transfers. In: 2016 IEEE International Conference on Cluster Computing (CLUSTER), Taipei, pp. 138–139 (2016). https://doi.org/10.1109/cluster.2016.74
28. Chawda, R.K., Thakur, G.: Big data and advanced analytics tools. In: 2016 Symposium on Colossal Data Analysis and Networking (CDAN), Indore, pp. 1–8 (2016). https://doi.org/10.1109/CDAN.2016.7570890

Decision Support System for Water Resources Management in the West Mediterranean Area-Morocco

Amal Sbai and Mohammed Karim Benhachmi[✉]

Department of Process Engineering and Environment,
Faculty of Sciences and Techniques of Mohammedia,
Hassan II University of Casablanca, Mohammedia, Morocco
Amalsbai95@gmail.com, benhachmikarim@gmail.com

Abstract. Sustainable management of water resources has become a major concern in many countries [2]. Morocco has implemented a water resources management policy that has allowed a relatively efficient mobilization of conventional, surface and underground resources available to it.

The main purpose of this study is to diagnose the area of action (West Mediterranean area) to compare the state of water resources over a period of 20 years based on geographic information systems (GIS) and remote sensing. To achieve this goal, the methodological approach adopted is summarized as follows: different data were collected and processed for the design of the maps (Land cover map, map of the different systems, map of water requirements …), These data will be subject to some geo-processing, this can be considered an addition for interpretation and decision-making.

The comparison results were transposed on digital maps to find a decrease in water resources due to several factors such as urbanization.

Keywords: Water resources · GIS · Remote sensing · Maps · Interpretation · Decision-making · Comparison

1 Introduction - Problematic

At the global scale, the problem of the rational management of water resources arises with the scarcity of this vital resource and the many factors of degradation [3]. Earth observation techniques and geographic information systems contribute to a better global and local knowledge of the distribution of these resources on the Earth's surface as well as a better understanding and analysis of hydrogeological phenomena. Several practical applications of these tools help managers and decision-makers with mapping, monitoring, inventory and management.

In fact, the rational management of water resources must make it possible to ensure, on a continuous basis, the availability of this commodity, sufficient in quantity and quality, for the benefit of all users, in accordance with the aspirations of harmonious economic and social development, sectoral plans and the technical and economic opportunities available.

M. Ezziyani (Ed.): AI2SD 2019, LNNS 92, pp. 81–87, 2020.
https://doi.org/10.1007/978-3-030-33103-0_9

For this reason, geographic information systems and remote sensing allow the integration of satellite images with exogenous data, facilitating information management and processing operations, modeling and simulation, and the production of data. Document synthesis [4].

This article aims at making the necessary maps to evaluate the contribution of remote sensing and Geographic Information System (GIS) in the management of water resources in the West Mediterranean zone in Morocco.

2   Study Area

The western Mediterranean zone is located in northern Morocco on an area of 3330 km^2 (Fig. 1), under mixed influence, includes the coastal area of Fnideq Oued El Had. It is mainly drained by wadis Smir, Martil, Amsa, Oued Laou and El Had.

The climate is humid in the Western Mediterranean Coastal Basin, which gradually decreases, induces an increasingly pronounced aridity from West to East to tend towards a Mediterranean-type climate with normal average annual temperatures vary between 14 °C and 20 °C with a minimum in January and a maximum in August. The water slide precipitated over the whole area is on average 910 mm. Rainfall maxima are generally between November and January. At least 65 to 80% of the annual rainfall totals are recorded during the period from November to March (LHBA) [1].

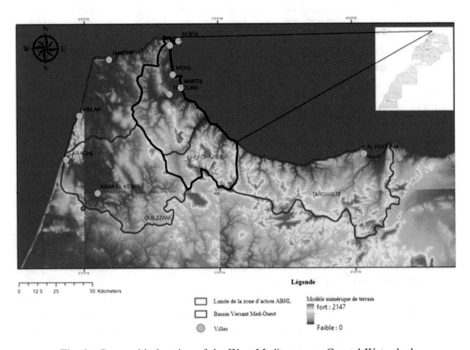

Fig. 1. Geographic location of the West Mediterranean Coastal Watershed

3   Materials Used and Methodology

3.1   Used Data

To achieve the desired objective during the realization of this article we used a 12.5 m resolution DEM, SENTINEL-2 satellite images and other auxiliary data such as the rainfall and the piezometric level of the aquifers. used to improve the methodology and have more information on the area studied, we opted for the choice of several software such as ArcGIS 10.5, ENVI 5.3 and Geoserver, PostGres, PostGIS and Openlayers for the web platform.

3.2   Methodology

In this article, the focus is on feeding the geodatabase, relying first of all on the delimitation of the watershed starting with the definition of our hydrographic network, then the determination of the watershed of each section of the water network [5]. flow, then the determination of the outflow of the main watercourse. And on the other hand the design and production of geological maps, vegetation, hydrographic network, distribution of rainfall, groundwater situation and distribution of interannual mean watershed contributions.

4   Results and Discussion

The results have been transposed on digital maps to transform them on the web platform.

4.1   Geology

Geologically, the area forms the westernmost part of the perimediterranean alpine belt, an orogenic segment that plays the strategic role of a crossroads linking Africa-Europe and the Atlantic-Mediterranean. The interactions between these great ensembles had built the arch of Gibraltar and all the mountain ranges surrounding the Alboran Sea, well known for their geological structures among the most complex on a planetary scale.

The predominant geological facies consist of limestones and karstic dolomites in the axial zone from Sebta to Al Hoceima (Fig. 2).

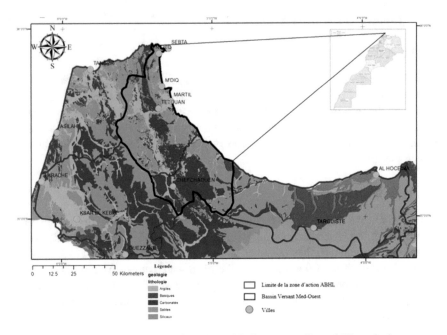

Fig. 2. Geological map of the West Mediterranean Coastal Watershed

4.2   The Vegetation

The vegetation cover (Fig. 3) is characterized by the presence of plant species that vary according to the nature of the soil and altitude. Thus, oak and cork are found on the Tingitane Peninsula, while holm oak people the central mountains of the Rif in association with the pine, itself relayed in altitude by the fir or cedar. This vegetation cover is constantly deteriorating because of the clearing which, combined with the uneven relief of the zone and the facies of the grounds, favor erosion, considered among the strongest of Morocco, with as a consequence the loss of the agricultural lands and the siltation of dam reservoirs.

4.3   Hydrographic Network

The hydrographic network (Fig. 4) from the coastal zone of Fnideq to Wadi El Had. It is mainly drained by wadis Smir, Martil, Amsa, Laou and El Had.

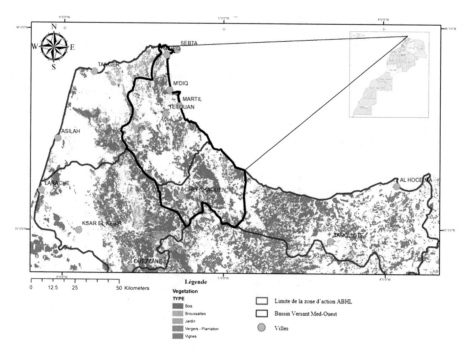

Fig. 3. Vegetation map of the West Mediterranean Coastal Watershed

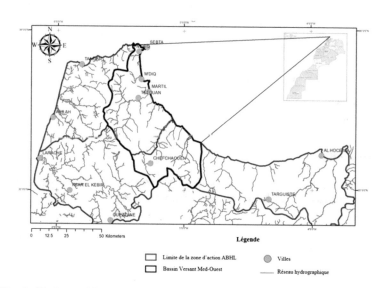

Fig. 4. Hydrographic network map of the West Mediterranean Coastal Watershed

4.4   Surface and Underground Waters

The West Mediterranean coastal basin has an abundance of surface water (Fig. 5), with per capita ratios of 1960 and 1880 m^3/yr/inhab respectively, almost double the water stress threshold of 1000 m^3/yr/inhab. Most of the inflow is recorded in the form of floods, as a result of the torrential hydrological regimes of the rivers in the area. This torrential nature is appreciated through peak flood flows that can reach several thousand m^3/s at the main rivers in the area.

The groundwater situation (Fig. 5) of the area by large homogeneous hydrologic unit is as follows:

Several coastal alluvial aquifers constituted in particular by the units of Martil-Allila, Oued Laou, Smir, Negro, Bouahmed, Amsa.

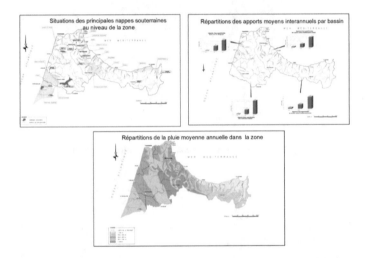

Fig. 5. Set of maps for the distribution of surface and groundwater in the Mediterranean West Coastal Watershed

4.5   Web Platform

The goal is to promote the development, promotion and harmonization of web geospatial standards and to have computerized maps (Fig. 6).

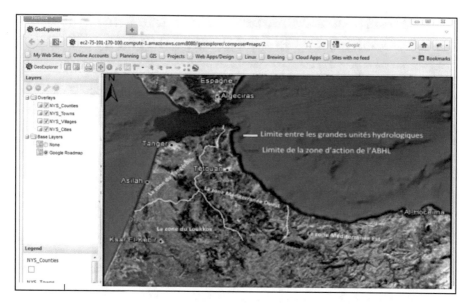

Fig. 6. Extract from the web platform

5 Conclusion

By way of conclusion, geographic information systems and remote sensing contribute to a better global and local knowledge of the distribution of resources on the Earth's surface and to a better understanding and analysis of the phenomena.

The results presented allowed the characterization of the Mediterranean coastal watershed, as well as the computerization of geographical data.

References

1. LHBA: Loukkos Hydraulic Basin Agency
2. Bonn, F.: Precision Remote Sensing. Thematic Applications, vol. 2, p. 633 (1996)
3. Googchild, M.F.: GIS and Environmental Modeling: Progress and Research, p. 486 (1996)
4. Abdelbaki, C., Benchaib, M.M., Benziada, S., Mahmoudi, H., Goosen, M.: Management of a water distribution network by coupling GIS and hydraulic modeling: a case study of Chetouane in Algeria. Appl. Water Sci. **7**(3), 1561–1967 (2017). https://doi.org/10.1007/s13201-016-0416-1
5. Lynn, E.J.: GIS and remote sensing applications in modern water resources engineering, p. 38 (2013). www.springer.com
6. Sauvagnargues-Lesage, S., Ayral, P.A.: Systèmes d'Information Géographique: outil d'aide à la gestion territoriale. Techniques de l'ingénieur, Référence, p. H7415 (2009)
7. Weng, Q: Remote sensing and GIS integration: theories, methods, and applications (2010)
8. Congalton, R.G., Green, K: Assessing the accuracy of remotely sensed data: principles and practices (2002)
9. Liu, J.G., Mason, P.J.: Essential image processing and GIS for remote sensing (2009)
10. Shahabi, H, Hashim, M: Landslide susceptibility mapping using GIS-based statistical models and Remote sensing data in tropical environment. Sci. Rep. (2015)

Design of a Miniature Microstrip Antenna with DGS Structure for RFID Tag

Mohamed Ihamji[1(✉)], El Hassan Abdelmounim[1], Jamal Zbitou[2], Hamid Bennis[3], and Mohamed Latrach[4]

[1] LASTI, FST, Hassan 1st University, Settat, Morocco
mihamji@gmail.com
[2] LMEET, FST, Hassan 1st University, Settat, Morocco
[3] TIM Research Team, EST, Moulay Ismail University, Meknes, Morocco
[4] Microwave Group, ESEO, Angers, France

Abstract. This paper presents a miniature microstrip antenna at 915 MHz in the ISM "Industrial Scientific and Medical" band. This microstrip antenna is designed for RFID tag system, using the DGS (Defected Ground Structure) and the slots technique, on FR4 substrate. The simulation is performed using CST Microwave Studio. The total area of the final circuit is 49.3×55.57 mm^2. The validated antenna has a good matching input impedance range from 909 MHz to 921 MHz with a stable radiation pattern, a loss return of -23.72 dB, a directivity of 1.85 dB, and a gain of 1.21 dBi.

Keywords: Microstrip antenna · RFID tag · Miniature antenna · DGS · Slots technique · 915 MHz · UCODE G2IL

1 Introduction

The demand of RFID systems with compact size has increased in many applications as access control, transport, bank, health and logistic. RFID system is composed from a Reader and one or more Tags [1].

The dimension of the tag antenna represents the most tag size, approximately 99% of tag size; therefore, the design of miniature tag antennas is very important to reduce the tag size. Many techniques are used to miniaturize the antenna size, as the slot technique, the fractal structure, the Defected Ground Structure (DGS), or the use of Metamaterial structure. Slots and DGS techniques are good candidates for miniature antennas, due to their simple structures [2–11].

The microstrip antenna is composed of three parts which are radiating element, substrate and ground. These antennas can have many forms such as rectangular, elliptic, circular and triangular, etc. They are famous to be low-profile, low weight, ease of fabrication, conformable to planar and non-planar surfaces and mechanically robust. Rectangular patch is the basic and the most used microstrip antenna in wireless communication [12, 13].

© Springer Nature Switzerland AG 2020
M. Ezziyyani (Ed.): AI2SD 2019, LNNS 92, pp. 88–99, 2020.
https://doi.org/10.1007/978-3-030-33103-0_10

Most RFID systems operate in either the low frequency region (30–300 kHz), the high frequency band (3–30 MHz), the ultra-high-frequency band (300 MHz–3 GHz), or in the microwave band (3 GHz–40 GHz) [1].

In this paper, a new design of a miniature low cost microstrip antenna for RFID tag is proposed at frequency of 915 MHz. This antenna is designed on a patch structure with symmetric slots inserted on the top side, and a DGS plane placed on the other side of the substrate. This circuit is validated by optimization methods integrated in CST-MW Studio [14–21].

2 Tag Antenna Design

The proposed antenna is composed of a rectangular radiating patch with a FR4 substrate that is characterized by:

- Dielectric constant $\varepsilon_r = 4.4$
- Loss tangent tan (δ) = 0.025
- Thickness H = 1.58 mm
- Metal thickness t = 0.035 mm

This antenna is designed for the UCODE G2IL chip from NXP Semiconductor which is used in small passive smart tags. At the 915 MHz frequency, the chip impedance is $23 - j\ 224\ \Omega$, The minimum threshold power of this RFID chip is -18 dBm (15.8 µW).

The antenna impedance must be equal to the conjugate of the chip impedance in order to have the maximum power transfer to the tag. There are many techniques used to obtain the matching impedance between the chip and the antenna as the nested slot, the T match, and the proximity coupled loop [22]. On this case, another matching technique is used; it consists to adjust the antenna geometry with simulation in order to have the matching impedance.

The size reduction of this microstrip antenna is obtained by inserting the slots into the patch antenna and replaced the ground plane by the DGS plane. The Defected Ground Structures (DGS) consist of slots or defects integrated on the ground plane of the antenna. The slots technique affects the resonant mode by changing the current distribution of patch.

The patch antenna has the length L and width W. It is connected to a microstrip line fed. The length L and width W are related to the resonant frequency, to the permittivity and to the thickness of substrate; this length and width can be calculated, theoretically for a conventional patch antenna, by the transmission line method [23].

As shown in Fig. 1, the patch has many slots inserted on the first side. A DGS plane is placed on the other side of the substrate, as is illustrated in Fig. 2.

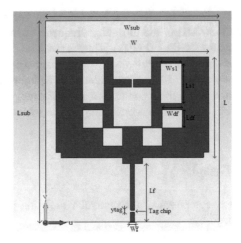

Fig. 1. Geometry of the proposed antenna: top face

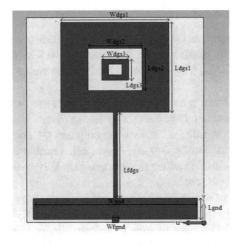

Fig. 2. Geometry of the proposed antenna: bottom face

3   Results and Discussion

The impedance matching is realized by adjusting the antenna geometries; many investigations have been done on some antenna parameters to understand their effects.

The first step, the position "df" of slots (Wdf, Ldf) is changed from 1.26 mm to 1.66 mm by a step size of 0.1 mm and all other parameters are kept constant. As can be seen in Figs. 3 and 4, the real and imaginary parts of antenna impedance increase when the slots position df is changed from 1.26 mm (near to patch low side) to 1.36 mm (far from the patch low side), and when the slots position is higher than 1.46 mm (more far from the patch low side).

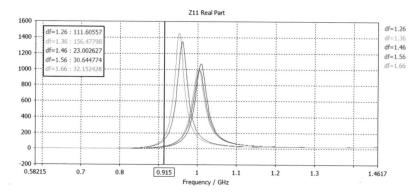

Fig. 3. Real part of antenna impedance for various "df" values

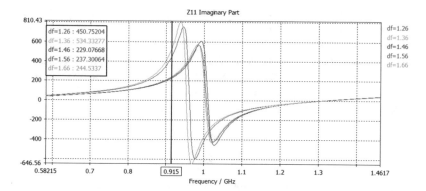

Fig. 4. Imaginary part of antenna impedance for various "df" values

Another parameter studied is the position of tag chip in the microstrip fed line "yvia", this parameter is varied from 1.8 mm to 3.8 mm by a step size of 1 mm. When the value of "yvia" increases, the real and imaginary parts of antenna impedance also decrease, as illustrated in Figs. 5 and 6.

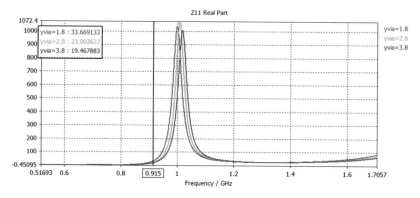

Fig. 5. Real part of antenna impedance for various "yvia" values

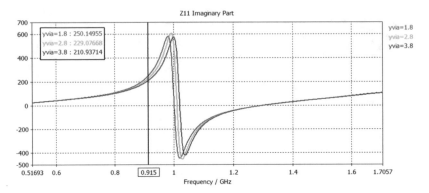

Fig. 6. Imaginary part of antenna impedance for various "yvia" values

Others parameters of antenna geometry (Wf, Lf) are also studied to show their effect. After optimization of antenna parameters, the antenna impedance is well matched with the chip impedance. The following Figs. 7 and 8 show the real and imaginary part of antenna impedance according to the frequency. The simulated antenna impedance obtained at 915 MHz is Zin = 23 + j 229 Ω, and therefore having good impedance match with the chip impedance which is 23 − j 224. The power transfer coefficient in this case is equal to:

$$\tau = \frac{4R_P.R_A}{|Z_P + Z_A|^2} = 0.988 \tag{1}$$

A good impedance matching could ensure a longer antenna reading distance, which is an important condition for RFID application.

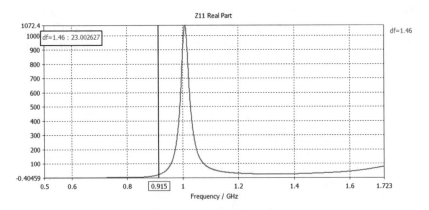

Fig. 7. Real part of simulated antenna impedance

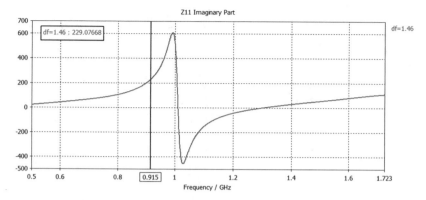

Fig. 8. Imaginary part of simulated antenna impedance

As explained before, the resonance frequency is related to the patch dimension (W, L), the slots size (Ls1, Ws1, Ldf, Wdf), and the DGS geometries (Wg, Lg). The following Fig. 9 illustrates the effect of the size of DGS plane. More the DGS size is higher (Wg is higher); more the resonance frequency is lower.

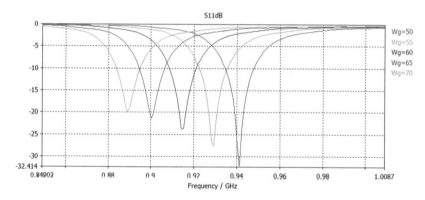

Fig. 9. Return loss S11 for various Wg values

Once the optimization of the antenna parameters is done, the antenna dimensions are adjusted to obtain the good matching at the frequency of 915 MHz. The following Table 1 indicates the dimensions of the proposed antenna:

Table 1. Antenna dimensions

Antenna top dimensions	Optimized value (mm)	Antenna bottom dimensions	Optimized value (mm)
Lsub	55.57	Lgnd	5.60
Wsub	49.30	Wgnd	45.70
L	27.77	Lfdgs	23.43
W	43.30	Ldgs1	24.26
Lf	16.09	Wdgs1	30
Wf	1.21	Ldgs2	11.50
Ls1	11	Wdgs2	15
Ws1	6	Ldgs3	5.75
Ldf	4.98	Wdgs3	7S.50
Wdf	4.62		
ytag	0.50		
yvia	2.8		

Figure 10 shows the return loss S11 result for the proposed antenna, it has good matching input impedance in the ISM band.

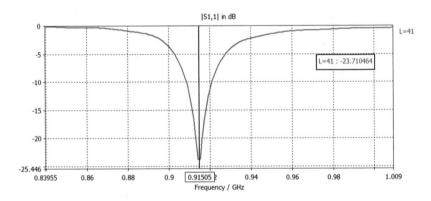

Fig. 10. Return loss (S11) value of the proposed antenna

The return loss (S11) obtained is equal to –23.72 dB at frequency 915 MHz. The bandwidth value is equal approximately to 12 MHz.

Figures 11 and 12 illustrate the 2D radiation patterns at E-plane and H-plane respectively. The proposed antenna has an omni-directional radiation pattern for H-plane, and a directional radiation pattern for E-plane; which means this antenna can be easily detected by reader antenna in any direction.

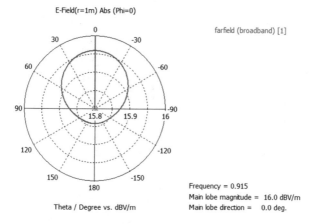

Fig. 11. 2D radiation patterns at E plane of the proposed antenna.

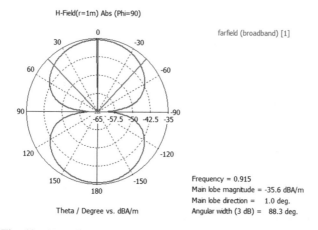

Fig. 12. 2D radiation patterns at H plane of the proposed antenna

Figure 13 illustrates the surface current distribution of the proposed antenna at 915 MHz. A maximum current is observed around the slots, the low side of DGS plane, and around the feed line.

The gain and directivity are equal respectively to 1.21 dBi and 1.85 dB, this gain is correct for RFID application in near field.

The reading range is a main feature of the RFID tag antenna; this parameter can be affected by a number of parameters such as, the gain, the frequency, and the power threshold to activate the tag chip. The maximum reading range can be computed by using the Friss Eq. (2):

$$R_{max} = \frac{c}{4\pi f} \sqrt{\frac{Eirp.\tau.G_t}{P_{th}}} \qquad (2)$$

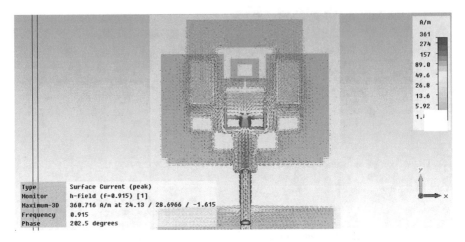

Fig. 13. Surface current distributions of the proposed antenna.

With:

- C: Speed of light = 3.10^8,
- f: Resonance frequency = 915 MHz,
- EIRP: Effective Isotropic Radiated Power. The value of this power is fixed by the Federal Communications Commission (FCC) to 36 dBm (4 W) at the frequency of 915 MHz,
- P_{th}: Power threshold to activate the tag chip = −18 dBm (15.8 µW),
- τ: Power transfer coefficient = 0.988,
- G_t: Gain of the tag antenna = 1.21.

The reading range of the proposed tag antenna, is estimated in this case, to Rmax = 14.4 m.

The results obtained of the proposed antenna, resumed in Tables 2 and 3, show that a good miniaturization of the antenna dimensions is obtained with a good return loss S11, and a good bandwidth for RFID applications. As shown in Table 2, the total size reduction compared to the conventional rectangular patch antenna is for example, in the case of the proposed antenna, equal to 72.1%. It is smaller than the True 3D antenna [24] (38%), and the Novel broadband UHF RFID tag antenna [25] (56.8%); It is slightly lower than the Miniaturized meandered line antenna [26] (71.9%).

The gain and the bandwidth of the proposed antenna are slightly lower than those of the meandered line antenna [26], but stay correct for the RFID application.

Table 2. Total size result comparison between proposed antenna and others antennas

Patch antenna 915 MHz	Wsub (mm)	Lsub (mm)	Total size reduction (Wsub × Lsub)
Conventional antenna	101	97.2	
True 3D antenna for UHF RFID application [24]	78	78	38%
A novel broadband UHF RFID tag antenna [25]	106	40	56.8%
Miniaturized meandered line antenna for UHF RFID tags [26]	77.68	35.5	71.9%
Proposed antenna	**49.30**	**55.57**	**72.1%**

Table 3. Result comparison between proposed antenna and others tag antennas

Patch antenna 915 MHz	S11 (dB)	Bandwidth (MHz)	Gain (dB)
True 3D antenna for [24]	−28.93	244	2.4
A novel broadband UHF RFID tag antenna [25]	−17.3	17	−
Miniaturized meandered line antenna for UHF RFID tags [26]	−	21	1.75
Proposed antenna	**−23.72**	**12**	**1.21**

4   Conclusion

This paper has presented a design of a new miniature tag antenna, with slots and DGS structures. A size reduction of antenna dimension (Wsub × Lsub) equal to 72.1% has been obtained compared with conventional rectangular patch. It provides appropriate characteristics, with a return loss equal to −23.72 dB, a bandwidth of 12 MHz, the directivity equals to 1.85 dB, and a gain equals to 1.21 dBi. This tag antenna has an omni-directional radiation pattern. The maximum reading range of the proposed tag antenna is estimated by 14.4 m at 4 W EIRP radiation power. The antenna has been designed on a standard FR4 substrate and will be realized with conventional Printed Circuit Board (PCB) techniques in order to validate its results. This antenna can be used in inventory control or asset tracking.

Acknowledgments. We thank Professor Mohamed Latrach for allowing us to use all the equipment and EM solvers available in his laboratory.

References

1. Finkenzeller, K.: RFID Handbook Fundamentals and Applications in Contactless Smart Cards, Radio Frequency Identification and Near-Field Communication, 3rd edn. Giesecke & Devrient GmbH, Munich (2010)

2. Ghiotto, A.: Tag UHF RFID antenna design, application by jet of material. Thesis, Institut Polytechnique Grenoble, France (2008)
3. Reynolds, N.D.: Long range ultra-high frequency (UHF) radio frequency identification (RFID) antenna design. Thesis, Faculty of Purdue University Fort Wayne, Indiana, US (2012)
4. Pigeon, M.: Design of ultra-compact antennas based on metamaterials, application to the fabrication of a miniature GNSS antenna. Thesis, Institut National Polytechnique Toulouse, France (2011)
5. Nithisopa, K., Nakasuwan, J., Songthanapitak, N., Anantrasirichai, N., Wakabayashi, T.: Design CPW fed slot antenna for wideband applications. PIERS Online 3(7), 1124–1127 (2007)
6. Deleruyelle, T., Pannier, P., Bergeret, E., Bourdel, S.: Dual band UHF and microwave RFID antenna. In: Proceedings of the 38th European Microwave Conference (2008)
7. Monti, G., Catarinucci, L., Tarricone, L.: Compact microstrip antenna for RFID applications. Prog. Electromagn. Res. Lett. 8, 191–199 (2009)
8. Reha, A., Elamri, A.: Fractal antennas: a novel miniaturization technique for wireless networks. Trans. Netw. Commun. 2(5), 166 (2014). Society for Science and Education, United Kingdom
9. Irfan, N., Yagoub, M.C.E., Hettak, K.: Design of a microstrip-line-fed inset patch antenna for RFID applications. IACSIT Int. J. Eng. Technol. 4(5), 558 (2012)
10. Anscy, S.: Slot microstrip antenna for 2.4 GHz RFID reader application. Int. J. Adv. Res. Electron. Commun. Eng. (IJARECE) 2(5) (2013)
11. Chandrappa, D.N., Ambresh, P.A., Vani, R.M., Hunagund, P.V.: Multi-slots reconfigurable microstrip antenna with capacitive loading technique. Int. J. Innov. Res. Comput. Commun. Eng. 2(1) (2014)
12. Nasimuddin, N.: Microstrip Antenna. InTech, Rijeka (2011)
13. Guha, D., Antar, Y.M.M. (eds.): Microstrip and Printed Antennas: New Trends, Techniques and Applications. Wiley (2011)
14. Ihamji, M., Abdelmounim, E., Bennis, H., Hefnawi, M., Latrach, M.: Design of compact tri-band fractal antenna for RFID readers. Int. J. Electr. Comput. Eng. (IJECE) 7(1) (2017)
15. Ihamji, M., Abdelmounim, E., Zbitou, J., Bennis, H., Latrach, M.: Novel design of a miniature fractal microstrip CPW fed antenna for RFID reader. In: IEEE Conference Publications, The International Conference on Wireless Networks and Mobile Communications (WINCOM) (2016)
16. Ihamji, M., Abdelmounim, E., Zbitou, J., Bennis, H., Latrach, M.: Novel design of a miniature L-slot microstrip CPW-fed Antenna for RFID reader. In: IEEE Conference Publications, The 11th International Conference on Intelligent Systems: Theories and Applications (SITA) (2016)
17. Ihamji, M., Abdelmounim, E., Bennis, H., Lotfi, M., Latrach, M.: A new design of a miniature L-slot microstrip antenna for RFID Tag. In: Proceedings of the 2nd International Conference on Computing and Wireless Communication Systems (ICCWCS) (2017)
18. Ihamji, M., Abdelmounim, E., Bennis, H., Hefnawi, M., Latrach, M.: A miniature L-slot microstrip printed antenna for RFID. TELKOMNIKA Ind. J. Electr. Eng. 16(5), 1923–1930 (2018)
19. Ihamji, M., Abdelmounim, E., Bennis, H., Latrach, M.: Design of miniaturized antenna for RFID applications. In: Emerging Innovations in Microwave and Antenna Engineering Book, pp. 325–362. IGI Global (2018)
20. Ihamji, M., Abdelmounim, E., Zbitou, J., Bennis, H., Latrach, M.: A compact miniature fractal planar antenna for RFID readers. Ind. J. Electr. Eng. Comput. Sci. (IJEECS) 15(1) (2019)

21. Ihamji, M., Abdelmounim, E., Bennis, H., Latrach, M.: Design of a miniature fractal microstrip antenna for RFID tag. In: Proceedings of the 3th International Conference on Computing and Wireless Communication Systems (ICCWCS19), EAI (2019). http://dx.doi.org/10.4108/eai.24-4-2019.2284079

22. Marrocco, G.: The art of UHF RFID antenna design: impedance-matching and size-reduction techniques. IEEE Antennas Propag. Mag. **50**(1), 66–79 (2008)

23. Fang, D.G.: Antenna Theory and Microstrip Antennas, p. 103. Taylor and Francis Group, LLC (2010)

24. Zhang, Y.S., Wang, B., Zhang, H.P., Kong, W.X.: True 3D antenna for UHF RFID application. In: Symposium on Quality Electronic Design, pp. 83–85 (2015)

25. Xu, L., Tian, L.B., Hu, B.J.: A novel broadband UHF RFID tag antenna mountable on metallic surface. In: International Conference on Wireless Communications, Networking & Mobile Computing, pp. 2128–2131 (2007)

26. Rokunuzzaman, Md., Islam, M.T., Rowe, W.S.T., Kibria, S., Singh, M.J., Misran, N.: Design of a miniaturized meandered line antenna for UHF RFID tags. PLoS ONE (2016). https://doi.org/10.1371/journal.pone.0161293

Distributed Spatio-Temporal Voronoi Diagrams: State of Art and Application to the Measurement of Spatial Accessibility in Urban Spaces

Hafssa Aggour[✉] and Aziz Mabrouk

ER Information Systems Engineering, Abdelmalek Essaadi University,
Route de Martil, Tétouan, Marocco
hafssaaggour@gmail.com, Aziz.mabrouk@gmail.com

Abstract. Irregular development and rapid changes are largely used to contribute to the production of large and uncontrollable data, making the management, analysis, processing, storage and interpretation of these massive spatial data extremely efficient. As a result, the displacement at the level of urban spaces becomes noticeably difficult. In this article, we are implementing a new approach that uses spatiotemporally voronoï diagrams based on a distributed architecture to solve large data processing problems on the one hand, and spatial accessibility in urban areas problems on the other hand.

Keywords: Big data · Voronoï network diagram · Distributed computing · Spatio-temporal modeling · Short path

1 Introduction

Currently, dynamic development in urban areas is increasing with an accelerated pace, due to demographic, economic, social and logistical growth. However, spatial accessibility in urban environments becomes very hard and more delicate. In addition, environmental changes and congestion create large data. Moreover, the problem that arises depends in particular on the analysis and the processing of the complex data which occur continuously in order to intervene quickly to the management and the development of the territory in terms of interrogation, simulation, implementation and planning on the one hand, and on the other hand to facilitate moving from one space to another in a calculated and optimal time.

Our contribution in this paper is to analyze spatiotemporal voronoï networks in a distributed way, in order to ensure a distributed processing of large spatial data produced. However, the voronoï diagram is a classical method that can model and analyze the space around a predetermined set of generators [25]. However, distributed computing is an effective solution that facilitates the processing of massive spatio-temporal data. The article highlights the different applications found in the literature to analyze voronoï spatial networks. Indeed, it clarifies the passage from the known approach that calculates spatial voronoï diagrams to a new approach that calculates spatiotemporal voronoï diagrams in a distributed way. Our article is composed of 5 sections. In the first

© Springer Nature Switzerland AG 2020
M. Ezziyyani (Ed.): AI2SD 2019, LNNS 92, pp. 100–117, 2020.
https://doi.org/10.1007/978-3-030-33103-0_11

section we present a general introduction to the theme in which we define the problem. In the second section we try to situate the theme in relation to other similar works with the representation of some application's fields of the theme and the technologies used in previous works. Before the conclusion, the 4th section presents the idea of our approach. It is an introduction to distributed spatiotemporal Voronoï diagrams and will be developed in future works.

This article is composed of 7 sections. At the level of the first section, we present a general introduction of the theme of which we define the problematic. The 2nd, 3rd, 4th and 5th sections represent the state of the art of the theme, at which we try to situate the theme in relation to other similar works. in the second section, we define the spatio-temporal modeling while trying to situate it in relation to other previous works, the 3rd section allows to present some fields of application of the distributed computation to facilitate the treatment of the giant data thus we present some work and technologies previously. in the 4th and 5th section we present work related to the computation of spatial network voronoï diagrams and the Application to the measurement of spatial accessibility in urban spaces and before the conclusion, we present our proposed approach in which we calculate the voronoï spatial-temporal network diagrams in a distributed way.

2 Spatio-Temporal Modeling

The spatio-temporal modeling of geographical areas is based on a set of concepts and methods that help to think of the modeling of space over time. However, fundamental spatial and temporal concepts and tools for studying geographic locations can be expressed in terms of locating and representing geographical features, describing their locations, method of studying network morphology and building regions homogeneous. Indeed, spatio-temporal analysis of geographical areas is mainly done by spatial modeling of these geographical areas expressed in a chronological interval.

2.1 Notion of Space and Time

Geographical phenomena are often decomposed into spatial objects (entities). An entity is characterized by its footprint in space [2]. On a Euclidean plane, space can be composed of points, lines, edges, arcs or polygons. The spatial imprint can be a surface but it can also be reduced to a point, a line [3–5]. The spatial relationships between entities are divided into three categories such as topological, metric and orientation. In addition, the temporal dimension is topologically similar to the spatial dimension [6]. Therefore, the study of the dimension of time is essential to better analyze the evolution of spatial entities and to study the phenomena considered. the reason which, Space and time are two dimensions usually inseparable from each other because the two complement each other. The author in [2] defines a spatio-temporal graph G_{ST} in the data model with is a pair (V, E): $V = \{O_i, \ldots \ldots, O_n\}$ The set of predicates relating to nodes such as $O_i \in O, \forall i \in [n]$. $E = \{R_1, \ldots, R_m\}$ all the predicates of edges such as for all $Ri \in G, \forall i \in [m]$, of the form Ri $(id_1, t_1, \rho, id_2, t_{2,})$, id_1, t_1, and id_2, t_2, are attributes in O_i and O_j, respectively, which are both in V.

2.2   Spatio-Temporal Relationship

In the field of geographic information, the two dimensions of time and space are perfectly indissociable [7], the spatio-temporal relation can be defined in the same way as the spatial relationship [2], this definition applies between entities whose times are different, but at the model level, it is essential to distinguish between the spatial relation whose entities are related to a given time and the spatio-temporal relationship. which expresses the way in which the entities are in relation in different times.

2.3   Related Works

Since its appearance, spatio-temporal modeling has been the subject of several scientific research in various fields. In fact, multiple works have been developed in this direction to show the effectiveness and importance of this modeling. Thus, in 2016, the authors in [8] tried to solve the problem related to the modeling of spatio-temporal data. Indeed, this modeling aims to design efficient solutions to process and analyze these data. In fact, these data are inscribed on spatio-temporal trajectories. As a result, the complexity of the phenomena and the size of the data make the representation of these trajectories in DBMS more difficult. Highlights, in order to design a spatio-temporal data model, [8] was used a database of an Origin-Destination (OD) survey, whose purpose of this survey is to establish a chart tracking the movements, behavior and transportation activities of each member of the selected sample. After designing the spatio-temporal data model, the author [8] has tried to implement this model by identifying significant temporal structures, incorporating these structures into the database schema and implementing predicates. 'Allen' as operations component time querier. Subsequently, the author [8] represented the selected survey results and road network data for the region in tabular form. In fact, the identification of significant temporal structures and the incorporation of these structures into the database schema make it possible to transform the schema of the static mode database into temporal mode by adding temporal attributes (Start - Time, End - Time, Duration) and the spatial attributes (From - Location, To - Location) to the objects of the database. These attributes serve to facilitate spatial and temporal computation for querying the database. Following the implementation of "Allen's" predicates, [8] used the Small world GIS shell with a spatial DBMS using a spatial data management engine to formulate and execute spatial, temporal, and spatial queries. spatio-temporal queries. On their part, the authors in [9] conducted a study to evolve the spatio-temporal interpolation method of wheat yield data, the main objective of which is to observe and record growing conditions of crops and weather environments by the stations. agrometeorological with greater pertinence and precision, to solve the problem of recording incomplete crop yield data in certain years. As a result, incomplete performance series data limit their application and cause drawbacks for information mining. Indeed, this study consists of reconstructing data in spatial and temporal dimensions based on spatio-temporal heterogeneity. Thus, the reconstructed results are combined with the back propagation neural network model for spatio-temporal integration. Subsequently, this study analyzes the key parameters and compares them with traditional interpolation methods. In the same vein, the authors in [10] proposed an approach to study the establishment of an adaptive spatio-temporal Gaussian filter whose filter facilitates the processing and analysis of Optical mapping of tissue types with low signal ratio. This study consists of showing how filtering parameters can be chosen automatically without additional user intervention. Of course, further in-depth analysis of the data requires the importance of digital filtering of optical fluorescence data. Therefore, although optical data mapping is the most widely used tool for studying cardiac electrophysiology, there is still a challenge in the analysis and processing of data collected from a thin mammalian myocardium.

3 Distributed Computing of Large Spatial Data (Related Works)

Massive production of semantic, spatial and temporal data in a continuous manner requires analysis and relevant processing. This processing is carried out using efficient algorithms, able to manage the complexities related to Big Data using distributed platforms of computation on several clusters so that the analysis is of an independent way whose purpose of acceleration processing time and decision making. Several technologies and works have been found in the literature and allow the use of distributed computing for to solve the problems of analysis and large-scale data processing. The table below gives an overview of some works, with the techniques used to develop them (Table 1).

Table 1. Theories used for distributed computing of large spatial data.

Authors/year	Implemented solutions	Method	Results
Zgaya [11]	Proposal a mobility assistance information system (Multimodal Transport Information System) based on multi-agent modeling to manage the daily, occasional movements in the urban environment	The TISM makes it possible to optimize the management of flows and the numerous and simultaneous queries expressed by the users whose multi-agent modeling ensures the system a continuous evolution and a pragmatic flexibility	This system makes it possible to draw powerful conclusions in terms of management and flow analysis of large road space data in a parallel way in order to ensure spatial accessibility in the urban environment
Bouha [12]	Proposal for a distributed multi-agent approach to improve traffic flow and optimize the flow of urban traffic	Driving rules and decision support tools based on the distributed multi-agent approach applied to urban road traffic systems optimize the flow of traffic	As a result, these distributed SMAs enable the analysis of massive data in order to optimize the movement of traffic at the urban level
Ahmed et al. [13]	Proposal of an approach named ASIF distributed for the calculation of CO2 emissions in the Moroccan logistic chains. The idea is to monitor CO2 production, aggregation of calculated rates and to estimate urban. The proposed distributed ASIF parameter collection model provides updated and parametric data by region, city, group of companies to ensure a synergy of work between the transport companies	In distributed treatment, the authors determine the energy intensity I from the energy consumption per unit of activity, and the carbon content F that should be communicated by the decision makers or extracted directly from the IPCC reports by region. the regional group of collaboration is modeled with a graph whose nodes represent the activity data (companies and emission Gi) and the modal structure communicate the energy intensity and the CO2 content	For validating the model, the experiment uses three types of means of transport (taxis, buses and intercity buses) to apply the ASIF model. As a result, the buses are more active and generate more activity with 61% of the modal structure updated, followed by intercity buses acting on the main points of the region with 34.87% followed by taxis. In short, although taxis are more numerous, they have a low occupancy capacity and generate less activity than buses and intercity

(continued)

Table 1. (*continued*)

Authors/year	Implemented solutions	Method	Results
Xia et al. [14]	Proposal of spatiotemporal weighted models distributed on MapReduce applied to road-based intelligent transport systems for the management and preventive control of short-term traffic flows	These models look for the nearest spatiotemporally K-weighted neighbor named STW-KNN. The latter is implemented on a Hadoop distributed computing platform adopted with the MapReduce parallel processing paradigm for real-time parallel traffic flow prediction	This approach helps to ensure more efficiency, effectiveness and relevance in order to accelerate traffic flows in a short-term, forward-looking way using STW-KNN distributed space-time weighted models
Boulmakoul et al. [15]	Proposal for a fuzzy e-pheromone-based proactive distributed system for predicting and analyzing congestion patterns of urban traffic in a cloud environment for large-scale data processing	This distributed system uses and the DBMS, NoSQL and MongoDB for storage and interaction of intelligent queries. This will be complemented by Hadoop which is designed to perform processing on large volumes of data. After processing, this data will be stored under an HDFS distributed file system	Hadoop ecosystem and private cloud virtual machines provide large data processing, where comparing this system with a physical server or the communication between MongoDB and Hadoop is done using the MongoDB-Hadoop-Connector to guarantee more than performance
Azedine Boulmakoul et al. [16]	Proposal of a distributed system based on IoT (Internet of Things) to analyze the dynamic spectral centrality of urban networks in real time. So that the variety of sensors they have the IoT capable of generating multiple nodes of massive data	Massive data is integrated into the processing pipeline for analysis and storage. The IOT uses Z-Wave and ZigBee lightweight protocols to send data to a central gateway that can aggregate data and integrate it into the system. The Apache Kafka cluster receives the data set from the gateway. The data that needs to be monitored and tracked in real time goes through the priority path	The result obtained is in the form of massive data analyzed under the Apache Kafka cluster and the IoT whose Apache Kafka deals with the processing of high-speed data ingestion but not with M2M. For this, the authors connect Kafka with MQTT to manage this ingestion of M2M
Kos et al. [17]	Proposal for a speech-based distributed architecture platform for a smart environment that allows the inclusion of multiple mobile robotic units for interaction between users and smart environments. Indeed, the proposed approach based on the architecture of the client/server platform integrates robust speech synthesis engines allowing a man-machine interaction in a more natural way between the users and the Genesis mobile unit. However, the Genesis Mobile Robotic Unit is an integral part of a distributed client/server architecture that includes proprietary speech recognition (ASR) and speech synthesis (TTS) synthesis engines that enable interaction between Genesis and users	In fact, the DATA platform comprises a main server, several module servers and one or more clients. The module server runs under Genesis, while the others must support client-side HMI interfaces to run ASR and TTS systems whose client software communicates directly with the primary server only. The primary server acts as a central management unit (MU), managing all module servers according to the user need. In addition, the main server and module servers can serve multiple clients in parallel. The primary server is responsible for establishing connections and exchanging data between all module servers and clients. the proposed system not only supports one mobile robotic platform, but also manages the flexible integration of additional mobile robotic platforms selected for use in a specific intelligent environment. The Genesis Platform Module Server is responsible for transferring video, audio, control and sensor data once the connection is established	As a result, the establishment of a distributed client/server architecture based on a human-machine interface (human machine interaction) requires the use of a mobile robotic platform called Genesis. Because, this mobile robotics platform (Genesis) can be hidden from the environment. As a result, users can benefit from the HMI interface anywhere. More than that, voice technologies used in HMI interfaces are heavy processing tasks that are best distributed. And this makes it easier for users to control and communicate with the mobile robotic platform

4   Spatial Network Voronoï Diagrams

4.1   Voronoï Diagrams

Space objects are modeled using points, lines and polygons. These primitives contribute to present the continuous phenomena [18]. Voronoï diagrams play a very important role in analyzing the space around a predetermined set of generators. These diagrams are based on the search for the generator of this diagram which is closest to each point in space.

4.2   Spatial Network Voronoï Diagrams

Two types of voronoï diagrams have been found in the literature, the first is called the planar voronoï diagram, which finds the nearest generator at each point in an E plane using Euclidean distance [19]. In fact, it is shown that the Euclidean distance is different from the distance of the shortest path in the urbanized area if the distance is less than 500 m [20]. This implies that urban spaces cannot be represented at using planar Voronoï diagrams. The reason why [20] uses another type of Voronoï diagrams is the Network Voronoï Diagram, replacing the plane with the Euclidean network and distance with the distance of the shortest way (Fig. 1).

Fig. 1. Part of spatial DVR generated by hospitals in the city of Tetouan (Source: LNCS, p. 7)

In 2008, the author [20] define a DVR by dividing the network into Voronoï subnetworks, each of which contains the points closest to each Voronoï generator while traversing the shortest path between these points and the nearest generators. If the analyzed network is a real spatial network (road network, transport network, etc.), this diagram is called the Spatial Network Voronoï diagram $(DVR_{Spatial})$.

$$Vor(i) = \{\forall p \in P / P(pi, p) \leq P(pj, p), 1 \leq \forall j \leq n, i \neq j\} \qquad (1)$$

4.3 Computing of Network Voronoï Diagrams

Several algorithms have been found in the literature to compute DVR citing for example, the algorithm A* and Dijkstra parallel algorithm. Indeed, the authors in [20, 21] used the Dijkstra parallel method to find the shortest path in a simultaneous way. Thus, [21] determines the efficient graph for determining the shortest path in a parallel way in order to find the nearest neighbor. Thus, he used two theorems, the first calculates the lower limit of the voronoï diagrams by the following complexity $\Omega(\max(n, (n-k)logk))$, the second uses the Dijkstra parallel algorithm to calculate the shortest path parallel with this complexity $O(m+nlogn)$. [20] divides the network into nodes and arcs whose diagram assigns each node and arc of the spatial network to a Voronoï generator such that the distance of the shortest path is the smallest compared to the others. Voronoï generators whose DVR computation time using Dijkstra parallel is $O(na+nslogns)$ with ns and na are respectively the number of nodes and the number of arcs constituting the network. On their part, the authors in [22, 23] used the A* algorithm to look for a shorter time-dependent path. In fact, the author in [22] used a KDXZ approach to path selection and time refinement. However, they maintain a priority queue Q among all the paths to develop in order to find the shortest path depends on the time. On his part, [23] tries to generalize the algorithm A * as an application, and also gives a generalization of the ALT algorithm for the classical problem, the shortest path depends on the time is calculated at using the following complexity:

$$Total\,cost\,(F) = Heuristic\,cost\,from\,source\,(G) + Cost\,to\,destination\,(H) \qquad (2)$$

4.4 Related Works

Voronoi Network diagrams are used extensively in many works. Indeed, the author in [21] used the Voronoi diagram to determine the nearest neighbor, closest installations and closest pair to ensure collision less motion and solve the problem of algorithmic geometry determination. He has put in place methods for structuring the data found in the databases storing the voronoï graphic information, then Erwig assumes that a graph G models the transport networks and that K is a set of nodes where facilities (fire stations, hospitals, postal centers or shopping centers, etc.) to illustrate the application of these voronoï diagrams. Finally, he deduced that these voronoï diagrams make it possible to calculate graphs G efficiently. However, M. Erwig distinguishes two algorithms for calculating the voronoï diagram, the first proves a lower limit of the problem, the second consists of identify the cases where the presented algorithms are optimal. For their part, [24] develop a spatial software system based on a fuzzy crisp network Voronoï diagram. In addition, several algorithms have been found in the literature to calculate the network-type Voronoï diagram but do not calculate the fuzzy network-type voronoï diagram. For this reason, [24] have extended this approach to compute network-type voronoï diagrams to compute the Voronoi network diagram.

$$Vor_{flou}(i) = \left\{ \forall p \in P/P_{flou}(p_i,p) \leq P_{flou}(p_j,p), 1 \leq \forall j \leq n, 1 \neq j \right\} \qquad (3)$$

In order to establish a fuzzy modeling of a real network (road network, transport network, etc.), it is essential to find the shortest fuzzy path to connect the nodes of the fuzzy space network to their nearest generators. of Voronoi (Schools, hospitals…). However, the author in [24] proposed a pseudo code of this method in which they used $R\,()$ a function of sorting numbers including:

$$Ppccflou(w) \geq \Delta flou \text{ if and only if } R\left(Ppccflou(w)\right) \geq R(\Delta flou) \qquad (4)$$

Indeed, [24] were inspired by the DVR-based spatial software system in order to develop a DVR-Fuzz based spatial software system. Whose software components of this Fuzzy software system are developed with these technologies (C++, ATL, ArcGIS Engine). However, the proposed Space System (SLSDV), which is an MFC application is developed under Visual C++, integrates a set of software components that can compute Voronoi crisp diagrams and evaluate spatial accessibility to places of interest. These software components integrated into the SLSDV, are then independent Component Object Model (COM) objects (in the form of Dynamic Link Library DLL) developed with Visual C++ and use the Active Template Library (ATL) as the COM technology was developed to allow inter-process communication.

5 Application to Spatial Accessibility Measurement in Urban Spaces

Voronoï diagrams are often used in many fields of application. Such as, natural sciences, health, computing, biology, logistics …. [25]. In addition, to recognize space around a predefined set of generators and solve several challenges related to Spatial accessibility at the level of urban spaces, multiple works have been elaborated in this direction and which make it possible to measure this spatial accessibility in urban environments by means of the spatial diagrams of voronoï [25].

5.1 Assessment of Accessibility and Proximity to Community Amenities According to Urban Standards

To assess the accessibility and proximity to community amenities in accordance with urban standards. The location, accessibility and population threshold to be served are the main criteria for the insertion of community-based amenities [25]. In fact, the author in [25] has deployed a spatial model object to objectively assess the situation of urban spaces by identifying the needs of the population and the dysfunctions in terms of service and accessibility. To do this, he used geographic maps as modeling tools to assess the spatial accessibility to the service areas using voronoï diagrams whose voronoï generators are the Tetouan city hospitals, as he represents the Fig. 2.

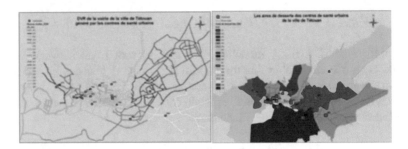

Fig. 2. DVR the city of Tetouan and service areas generated by urban health centers (Source: LNCS, p. 10)

Emphasizes, the spatial accessibility assessment does not only depend on the proximity to community facilities but depends on the distance and number of the population. The author in [25] found that the measure of spatial accessibility of the population served to hospitals is determined by the distance between a point and the health center.

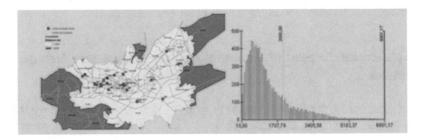

Fig. 3. Remote spatial accessibility to urban health centers in the city of Tetouan (Source: LNCS, p. 11)

Or according to a range of the population not to be exceeded. In this way, each health center must contain a very specific number of the population.

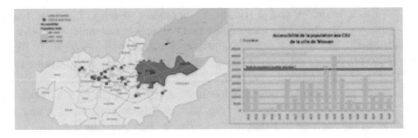

Fig. 4. Spatial accessibility of the population to urban health centers in the city of Tetouan (Source: LNCS, p. 11)

So, if the population threshold is exceeded, urban planning experts must make the decision to build a new health center to meet the needs of this population. Figures 3 and 4 measure respectively the spatial accessibility of the population and the distance to the urban health centers of the city of Tetouan.

5.2 Assessment of Urban Vulnerability Based on a Spatial Model of Voronoï

In another work, the author in [25] proposed to measure spatial accessibility by assessing spatial vulnerability in relation to proximity to accident zones in urban areas. For example, rapid and effective emergency response within a timeframe optimized by the evacuation of accident zones can reduce the real consequences of an accident [25]. In fact, he considered accident zones as generators of polygonal Voronoï, whose space he has cut into several zones in order to determine the targets closest to each accident zone. Subsequently, he calculated the Euclidean distances according to their spatial references (Fig. 5).

Fig. 5. Assessment of spatial accessibility to accident zones and hospitals in the city of Tetouan (Source: LNCS, p. 31)

The spatial vulnerability assessment cannot be calculated without estimating the response time of the rescue. However, the estimation of this delay depends mainly on the notion of risk analysis and influence of the emergency services [26]. Indeed, the author in [26] has proposed a calculation process that allows the calculation of Euclidean distances and associates the new information to all nodes constituting the roads to generate the Voronoï Network Diagram. In addition, the Voronoï Network Diagram (DVR) is used to calculate spatial accessibility to a set of places of interest. [26] determined the generators of Voronoï (emergency services, hospitals, …) closest to each set of nodes (accident zones) of the spatial network. By specifying the distance of the shortest path between the spatial components (voronoï generators and accident zones), the response time is decreasing [26] is therefore the spatial vulnerability is decreasing.

5.3 Planning for Hazardous Materials Transportation Routing in an Urban Environment

Hazardous material transport routing planning minimizes the risk of accidents [26]. This planning is based on the search for the safest and shortest roads based on the Voronoï graphs. Hence, the main idea is to browse the nodes furthest away from the Voronoï generators which these generators represent the risk targets. Dangerous goods vehicles must keep the greatest distance by traveling the shortest road to the destination. Figure 6 shows the safest and shortest roads.

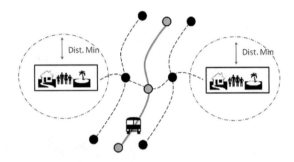

Fig. 6. The safest routes the shortest (Source: LNCS, p. 34)

In order to determine the minimum distance of vulnerable targets [26]. generates in parallel nodes called "division points". These points are found at the same distance from two or more voronoï generators. These nodes are furthest away from other nodes of each Voronoï subnet and they are not directly connected. Afterwards, the author in [26] looks for the shortest path between each two division points in order to construct a graph composed of these division points that are connected by a spatial adjacency relation. Hence, it gets the shortest and furthest road where TDG vehicles have to travel to reach their destinations.

The computing process proposed by [26]. is a rectification of the Dijkstra algorithm based on the result provided by the voronoï diagram and which results in a graph $G_{Vor} = (V_{Vor}, E_{Vor})$ where V_{Vor} is a finite set of nodes whose distance $dist_G[v]$ is available and which represents the distance of the shortest path connecting each node v to a generator of Voronoï G. This algorithm then makes it possible to find the shortest paths traveled by a TMD vehicle of a node data s (the starting point) to all other nodes $v \in V$ (the destination points).

As a result, if $Dist_G[V] > Dist_{Min}$ is the minimum distance, the TMD vehicle must keep bypassing a voronoï G generator. The advantage of this algorithm is that it is closer to the backup and rescue services. emergency and at the same time transport vehicles keep a minimum distance from vulnerable targets. Figure 7 shows the minimum distance of vulnerable targets (Fig. 7).

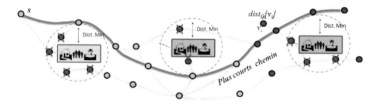

Fig. 7. Minimum distance of vulnerable targets (Source: LNCS)

6 Perspective: Distributed Spatio-Temporal Voronoï Diagram

In an environment full of vitality, activities and changes. The evaluation of spatial accessibility becomes very difficult. In fact, the spatial network voronoï diagram is a solution found in the literature that is designed to model and analyze spatial networks. [26] and therefore, they ensure spatial accessibility in urban spaces. On the real plane, this type of diagram is limited, since it allows to analyze just spatial networks and neglects the dimension of time, especially that the dimensions of space and time consider themselves indissociable between them [7]. In addition, the dimension of time is indispensable in the calculation and analysis of space-time networks, it helps in the decision-making of multiple phenomena. The analysis of spatio-temporal voronoï networks involves monitoring this network as a function of time, with the main goal being to manage spatio-temporal phenomena. For example, monitoring the transport routes of hazardous materials as a function of time minimizes the risk of accidents related to this type of vehicle. Another example, the road tracking of an ambulance over time gives an idea of the time consumed to arrive at a hospital. Cost, the time consumed to arrive at the hospital impacts the health status of patients especially if it is an urgent case that requires a rapid intervention.

The voronoï spatial networks carry an indeterminate number of spatial data such that their processing is not valid on a conventional analysis platform (relational databases ... etc.) and absolutely require processing on a distributed platform. In fact, the generation of Network Voronoï Diagrams depends mainly on the calculation of a shorter path. These paths carry a lot of data. Therefore, it is essential to analyze these voluminous spatial data to be able to calculate the shortest path and generate the Network Voronoï Diagram. On the other hand, existing Network Voronoï Diagrams in the literature can be used to analyze just the space while the temporal dimension is also important. For this, we propose a new approach to compute spatio-temporal voronoï diagrams based on a distributed architecture which this type of computation allows to ensure the processing, the storage, the management, the analysis, the interpretation and the display of large spatial data that continuously occurs efficiently. What are distributed processing technologies capable of analyzing large space-time data? how can one compute the spatio-temporal Network Voronoï Diagram?

6.1   Distributed Spatio-Temporal Voronoï Diagram

In order to solve the problem related to the computation of massive spatio-temporal data, our contribution in this article aims to develop a process of computation of Voronoï spatio-temporal networks based on a distributed architecture adapted to the context and the objectives of the study of real geographical phenomena. The use of distributed computing and the integration of the spatio-temporal dimension remain one of the current challenges in geographic information science research. Indeed, this process will allow a fast and reliable qualitative and quantitative processing by taking advantage of the distributed architecture for the acceleration of the calculations as well as the optimization of the processing time and interpretation of the massive data.

6.2   Representation of Spatio-Temporal Data of the Network in a Distributed Platform

Let $R(N, E, f(t))$ be a spatio-temporal network of which N is the set of nodes $N = \{n1, \ldots, nk\}$, E is the set of edges $E = \{e1, \ldots, eq\}$ and $f(t)$ the time dependent function to follow the network path in a chronological time. Let G be the set of m nodes representing the voronoï generators so that $G = \{g1, \ldots, gm\}$ with $G \, \varepsilon \, N$ (Fig. 8).

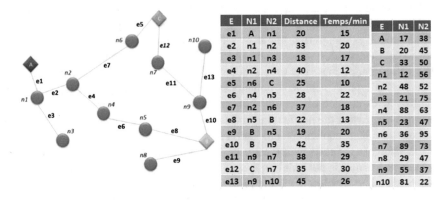

E	N1	N2	Distance	Temps/min
e1	A	n1	20	15
e2	n1	n2	33	20
e3	n1	n3	18	17
e4	n2	n4	40	12
e5	n6	C	25	10
e6	n4	n5	28	22
e7	n2	n6	37	18
e8	n5	B	22	13
e9	B	n5	19	20
e10	B	n9	42	35
e11	n9	n7	38	29
e12	C	n7	35	30
e13	n9	n10	45	26

E	N1	N2
A	17	38
B	20	45
C	33	50
n1	12	56
n2	48	52
n3	21	75
n4	88	63
n5	23	47
n6	36	95
n7	89	73
n8	29	47
n9	55	37
n10	81	22

Fig. 8. Modeling of the spatio-temporal network in graph (Source: LNCS)

Each edge e_j is valued by a weight W_{ej} that is to say each arc must represent a real number. Each node nor including the voronoï generator g_j are spatially referenced respectively with the coordinates $n_i(Xn_i, Yn_i)$ and $g_j(X \, G_i, YG_i)$, the function of the time $f(t)$ and which makes it possible to follow the road in a chronological time is represented by one of the units of measure of time. Nodes and arcs and the function of time are loaded as a set of resilient distributed data (RDD), the basic abstraction in a distributed platform. It represents a partitioned collection of elements that can be manipulated in parallel.

6.3   Partitionnement du réseau spatio-temporel de Voronoï

The computation process receives the spatial and temporal data of the network in the form of a set of resilient distributed space-time data (space-time RDD) for which these network data must be partitioned. Subsequently, the process proceeds to the distributed computation of the trees of the shortest time-dependent paths in order to generate the spatiotemporal voronoï diagram in a distributed manner. However, the idea is to partition the data of the space-time network according to the Euclidean proximity of the nodes with respect to the voronoï generators. In fact, the calculation process that we proposed divides the spatio-temporal network into n_g (the number of voronoï generators) subnets as a function of time. Each sub-network *SubNet* is composed of a sub-network of nodes *NSubNet* and a subnet of arcs *E_SubNeti* with (Fig. 9):

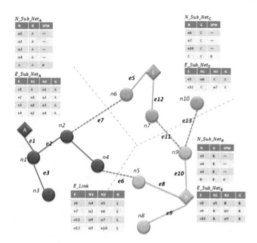

Fig. 9. Spatio-temporal partitioning of the network based on the distributed calculation of the Euclidean distance (Source: LNCS)

6.4   Distributed Calculation of Trees Shorter Path Depends on Time

The spatio-temporal data are divided into n_g partners. Now these data are ready to load in n_g gexecutors for parallel processing. Each partition j is a tree of shorter path depends on the time whose calculation process follows the parallel Dijkstra algorithm to look for the nodes closest to the root g_j (voronoï generator) of the tree *SubNetj* based on the weight of the paths traveled (Fig. 10). It marks the predecessor $Pr(n_i)$ and the weight SPW $[n_i]$ of the shortest path to reach each node n_i. The tree of the shortest paths $ACC(g_j)$ is then constructed by adjacency lists. The result is a set of trees of the shortest time-dependent paths $ACC(g_j)$ rooted at $g_i, i = 1..., n_g$.

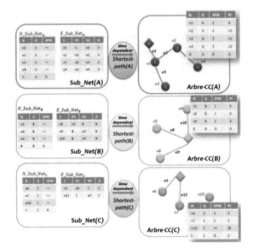

Fig. 10. Distributed computation of the trees of the shortest time-dependent paths rooted to the voronoï generators *A, B et C*

6.5   Update Generator Assignments to Vertices and Edges

The trees of the shortest paths *ACC* (*gi*) rooted at *gi*, *i* = 1 …, *ng*; are computed in parallel and in an independent way in which each *ACC* (*gj*) comprises all the points of reference points, by constituting a graph where each vertex contains the information on the weight of the shortest path to reach the root g_i in terms of time. However, this weight may be greater than the weight of the shortest path to reach another Voronoi generator. That said, updating the vertex assignment to the nearest generators is necessary. In fact, each two vertices constituting the ends of the connecting arcs do not belong to the same *ACC* tree. Consequently, it is the vertices with which we must begin the comparison of the weights of their shorter paths [27] (Fig. 11).

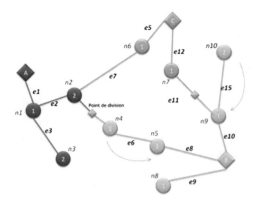

Fig. 11. Update generator assignments to vertices and edges

6.6   Technologies for the Implementation of the Proposed Process

The processing process we proposed begins with data collection (spatial data, distance, time, fuel consumption, etc.). These data are then bulky and do not admit conventional treatment. Thus, the distributed treatment remains the most adequate. The implementation of the process will be passed under the Apache Spark Framework whose development of the program is at the level of its native Scala language. Since this is parallel processing, Apache Spark with SBT takes care of the partitioning of the data (of the road network) whose program itself seeks to find the shortest path depends on the time via the parallel Dijkstra algorithm for several nodes whose nodes are the voronoï generators. Towards the end the calculation process will generate the distributed spatio-temporal voronoï diagram. This generated diagram will allow us to evaluate spatial accessibility in urban spaces (smart cities).

7   Conclusion

Parallel or part processing of massive spatial data is currently a priority in the continuous production of data (text, graph, etc.). Indeed, among the requirements of the spatial analysis, the establishment of a calculation process that allows to compute spatiotemporal diagrams of distributed voronoï. So, real-time data processing makes it possible to analyze and make fast, spatio-temporal decisions. This article is a literature review of voronoï diagrams and their applications in urban spaces, spatio-temporal modeling and distributed computing of large spatial data. In fact, it is an initial presentation and an introduction to our approach that we will develop in a future work and trying to prepare results showing the contribution of DVR Spatio-temporal distributed compared to other methods in terms of efficiency, execution time and adaptability to the types of data used and the situations modeled.

The perspective in this work is to implementation of this computational process with a Spark distributed platform that offers a comprehensive and unified framework to meet the needs of Big Data processing for various datasets, various by their nature (text, graph, etc.) as well as by the source type (batch or real-time stream). We exploit this distributed process in a future work to spatially analyze smart cities, taking advantage of its ability to geo-process massive spatial data.

References

1. Mabrouk, A., Boulmakoul, A., Karim, L., Lbath, A.: Safest and shortest itineraries for transporting hazardous materials using split points of voronoï spatial diagrams based on spatial modeling of vulnerable zones. Procedia Comput. Sci. **109**, 156–163 (2017). https://doi.org/10.1016/j.procs.2017.05.311
2. Mondo, D.G.: Un modèle de graphe spatio-temporel pour représenter l'évolution d'entités géographiques, thèse, l'UBO, pp. 1–163, octobre 2011
3. Kurata, Y., Egenhofer, M.J.: The head-body-tail intersection for spatial relations between directed line segments. In: GIScience Geographic Information Science. Lecture Notes in Computer Science, vol. 4197, pp. 269–286 (2006)

4. Freksa, C.: Using orientation information for qualitative spatial reasoning. In: Spatio-Temporal Reasoning, pp. 162–178 (1992)
5. Schlieder, C.: Reasoning about ordering. In: Frank, A., Kuhn, W.E. (eds.) COSIT, Lecture Notes in Computer Science, vol. 988, pp. 341–349 (1995)
6. Langran, G., Chrisman, N.R.: A framework for temporal geaographic information systems. Cartographica 25(3), 1–14 (1988)
7. Parkes, D., Thrift, N.: Times, spaces and places. a chronogeographic perspective. Geograph. J. 147(2), 247 (1981). https://doi.org/10.2307/634576
8. Frihida, A.: Contribution à la modélisation des données spatio-temporelles. Traitement du signal et de l'image. Ecole Nationale d'Ingénieurs de Tunis, pp. 13–20 (2015)
9. Jiang, L., et al.: A neural network method for the reconstruction of winter wheat yield series based on spatio-temporal heterogeneity. Comput. Electron. Agric. 154, 46–53 (2018). https://doi.org/10.1016/j.compag.2018.08.047
10. Pollnow, S., et al.: An adaptive spatio-temporal gaussian filter for processing cardiac optical mapping data. Comput. Biol. Med. 102, 267–277 (2018). https://doi.org/10.1016/j.compbiomed.2018.05.029
11. Zgaya, H.: Conception et optimisation distribuée d'un système d'information d'aide à la mobilité urbaine: Une approche multi-agent pour la recherche et la composition des services liés au transport, p. 241 (2007)
12. Bouha, N.: Modélisation, approche multi agent et outils d'aide à la décision applique sur les systèmes du trafic routier urbain, p. 8 (2014)
13. Ahmed, H.S., et al.: Une approche ASIF distribuée pour le calcul des émissions CO2 dans les chaines logistiques marocaines (2014). https://doi.org/10.13140/rg.2.1.2615.0166
14. Xia, D., Wang, B., Li, Y., Rong, Z., Zhang, Z.: An efficient MapReduce-based parallel clustering algorithm for distributed traffic subarea division. Discrete Dyn., Nat (2015)
15. Boulmakoul, A., Karim, L., Daissaoui, A.: Un système distribué proactif à base d'e-phéromone floue pour la prévision et l'analytique des patterns de congestions du trafic urbain dans un environnement cloud, p. 15 (2016)
16. Mabrouk, A., Boulmakoul, A.: Nouvelle approche basée sur le calcul des itinéraires courts et sûrs pour le TMD favorisant l'accès rapide aux secours, pp. 7–9 (2017)
17. Kos, M., et al.: A Speech-based distributed architecture platform for an intelligent ambience. Comput. Electr. Eng. 71, 818–832 (2018)
18. Ledoux, H., Gold, C.M.: Modelling three-dimensional geoscientific fields with the voronoï diagram and its dual (2006)
19. Okabe, A., Boots, B., Sugihara, K., Nok Chiu, S., Kendall, D.G.: Spatial Tessellation: Concepts and Applications of Voronoï Diagrams. Wiley, Chichester (2000). https://doi.org/10.1002/9780470317013
20. Okabe, A., Satoh, T., Furuta, T., Suzuki, A., Okano, K.: Generalized network Voronoï diagrams: concepts, computational methods, and applications. Int. J. Geograph. Inf. Sci. (2008)
21. Erwig, M.: The graph Voronoï diagram with applications. Networks, 156–163 (2000)
22. Kanoulas, E., Du, Y., Xia, T., Zhang, D.: Finding fastest paths on a road network with speed patterns. In: ICDE, pp. 10–19 (2006)
23. Zhao, L., Ohshima, T., Nagamochi, H.: A* Algorithm for the time-dependent shortest path problem, pp. 3–6 (2008)
24. Mabrouk, A., Boulmakoul, A.: Logicielle intégrant un système spatial décisionnel pour la géo-gouvernance des réseaux urbains. In: Conference ASD, pp. 8–9 (2013)
25. Mabrouk, A., Boulmakoul, A.: Modèle Spatial Objet basé sur les Diagrammes Spatiaux de Voronoï pour la géo-gouvernance des espaces urbains. In: Conference INTIS, pp. 10–11 (2012)

26. Mabrouk, A., Boulmakoul, A., Karim, L.: Support d'aide à la décision basé sur un modèle spatial de Voronoï pour la géo-gouvernance des réseaux de transport de matières dangereuses dans un milieu urbain. In: Conference INTIS 2016, p. 14 (2016)
27. Mabrouk, A., Aggour, H., Boulmakoul, A.: Processus de calcul parallèle des réseaux spatiaux de Voronoï basé sur une architecture distribuée. In: Conference ASD, pp. 2–13 (2018)

Elector Relationship Management: Concepts, Practices and Technological Support

Jalal Boussaid$^{(\boxtimes)}$ and Hassan Azdimousa

Ibn Tofail University, Kenitra, Morocco
{jalal.boussaid, hassan.azdimousa}@encgk.ma

Abstract. Election abstention is generally high during political campaigns and it is widely recognized that one of the ways to promote successful elections is to put in place mechanisms to track the assessment of their success.

To support electoral campaigns, it is essential to gain knowledge about citizen-electors. This knowledge will enable the adoption of adequate and effective actions and decisions to closely monitor elector behavior. For such procedures to be possible, this paper proposes an Elector Relationship Management (ERM) system. This system will support the ERM concept and practices and will be implemented using the concepts and technology infrastructure supporting Business Intelligence systems. The concept, practice and architecture of the ERM system is presented in this article and its main purpose is to provide a technological tool that helps political parties to acquire the essential knowledge to the decision-making process. The prototype of the ERM system proposed, once implemented, will be validated by the execution of a set of demonstration cases in different political parties in Morocco.

Keywords: Costumer Relationship Management · Business Intelligence · Elector Relationship Management (ERM) · Data warehousing · Data mining

1 Introduction

During election campaigns, abstention from voting is generally high. Since the 1980s, many political parties around the world had to cope with declining membership and reduced citizen participation [1]. Various studies were developed to identify and analyze the reasons for this abstention, as well as to suggest ways to promote the success of an election campaign [2]. The set of problems posed by failure is studied and analyzed in different ways [3, 4]. It is widely recognized that one of the ways to promote successful elections is to put in place mechanisms to track the assessment of their success. This involves, among other things, the development of processes to track citizen-electors and monitor their behavior. Although many efforts have been made, these processes do not take place in a large number of political parties due to a lack of resources, institutional practices and technological support to closely monitor citizen-electors. In addition, no technological tool available on the market properly supports these practices.

In order to overcome these conceptual and technological limitations, this paper proposes a system to help political parties closely monitor their citizen-electors and

© Springer Nature Switzerland AG 2020
M. Ezziyyani (Ed.): AI2SD 2019, LNNS 92, pp. 118–126, 2020.
https://doi.org/10.1007/978-3-030-33103-0_12

their electoral behavior. The proposed system will be designated by the ERM system - Elector Relationship Management System - and takes into account the concept of ERM and the practice of ERM, which will also be defined in this paper. Underlying to the ERM concept is the electoral success promotion, through the maintenance of a closely relationship between the political party and the citizen-elector. To this end, a set of activities that integrate the ERM practice will be defined.

This paper is organized as follows: the first section presents the motivation and justification for the development of the ERM system; its concept and practice, the second section describes the technological support provided to the practice of ERM, the proposal of an ERM system and its implementation.

2 Elector Relationship Management: Concepts and Practices

2.1 From Customer Relationship Management (CRM) to Elector Relationship Management (ERM)

This section describes the main differences between CRM and ERM because it is quite difficult to separate them and can be considered as confusing. First, it is important to clarify the meaning of both concepts in order to better understand their differences. According to Schellong, CRM is "a holistic management approach, based on customer-centric technology that creates, maintains and optimizes relationships and builds customer loyalty/profitability [5], while we define ERM as a set of management practices, communication channels and technological solutions to address the issues, concerns and demands of the citizen-electors by the political parties. As can be seen, Table 1 illustrates the main similarities/differences between CRM and ERM:

Table 1. Differences between Customer Relationship Management and Elector Relationship Management (Source: Adapted from [5, 6])

	Customer Relationship Management	Elector Relationship Management
Abbreviation	CRM	ERM
Field	Private sector	Political sector
Strategy	Companies	Political Parties
Organization based on	Competition	Competition
Goal	Optimize customer long-term value within the customer lifecycle Increase sales	Improve citizen-elector orientation Better accountability Modify the citizen-elector-political parties relationship
Focus	Customer needs	Citizens-electors needs
Foster	Effective marketing Sales Services processes	Transparency Citizen-elector feedback Opportunities to participate

(continued)

Table 1. (*continued*)

	Customer Relationship Management	Elector Relationship Management
Concern	Intense competition No profits Customer disloyalty Costs	Electoral abstention Complaints from citizens Insufficient communication and accountability with citizens-electors
Channels of Communication	Shop, outlets, counter, Sales force Internet, Mobile, ICT	Internet, Mobile, ICT, television media, general meeting, press conference, face-to-face
Measures	Quality Performance	Information Participation
Outcome	Profit Customer loyalty Maximising the shareholder value	Close citizen-elector relationship strengthening democracy and legitimacy Citizens-electors reactions and participation

Nevertheless, it is relevant to say that the ERM is obtained from the concept of global strategy of CRM with the customers [7]. Practitioners pay considerable attention to the transfer of concepts from the private sector to the public sector [5], so we think it is appropriate to also pay particular attention to the transfer of these concepts to the political sphere.

Thus, organizations need knowledge about their customers so they can develop a relationship with them that includes values such as trust, loyalty and sustainability. To do this, organizations need to develop specific strategies and leverage their internal capabilities and skills. The quality of customer relations is a factor of great importance because they should allow the company to stand out from the competition in relation to its customers [8].

Based on these elements, the organization must define and implement organizational practices and specific actions to maintain a close relationship with the client [9]. As mentioned earlier, the ERM system was inspired by the principles of CRM, but supports the processes and activities related to election campaigns, with the main purpose of reducing electoral abstention. As a result, the ERM system must be based on the definition of a concept, the ERM concept, and a set of activities that support this concept, the ERM practices. These fundamental concepts will be presented in the next subsection.

2.2   Implementation of an ERM System

In the private sphere, the implementation of the CRM system is commonly used in functional areas such as customer support and services, sales and marketing [10]. The CRM lifecycle consists on three stages: Integration, Analysis, and Action [11]. In the first step, the life cycle of CRM begins with the integration of front office systems and the centralization of customer data [12]. The second step called analysis is the most important for the success of CRM [10]. Analytical CRM enables effective customer relationship management [12]. By using analytical CRM, organizations are able to analyze customer behavior, identify customer buying patterns, and discover casual relationships [10]. The final phase, Action, is when strategy and decisions are executed. Business processes and organizational structures are refined based on improved customer understanding gained through analysis [13]. This step closes the CRM loop and allows organizations to take advantage of valuable information gained through analysis. Systemic approaches to CRM enable organizations to effectively coordinate and maintain the growth of different customers, points of contact or communication channels. The systems approach puts CRM at the heart of the organization, with customer-centric business processes and CRM systems integration [14].

In the political sphere, given that there is a strong correlation between the close monitoring of citizen-elector behavior and the success of an election campaign, the concept of ERM is understood as a process based on the behavior of citizen-electors, whose main objective is to maintain a close and effective relationship between its citizen-electors and political parties by closely monitoring their political choices throughout an election campaign. The practice of ERM is understood as a set of activities, defined by the political party, which must guarantee a personalized contact with the citizen-voters and an effective, adequate and detailed follow-up of their behaviors. The ERM strategy is understood as a means of defining the main objectives underlying the concept and practice of managing the citizen-elector relationship. This strategy must be aligned with the vision, mission and objectives of the political party and involve all persons in charge of responsibilities within that party.

The political party adopting an ERM initiative must define:

- An ERM strategy that engages the party in this practice, including the activities to be developed in this area, as well as the different participants in this practice.
- The identification of the performance indicators and behaviors that characterize the different situations to be supervised.
- The processes of monitoring the citizen-elector and his behavior,
- The validation of the ERM practice, using the results obtained and, if necessary, the redefinition of the practice.

3   Proposal for an ERM System

3.1   The Importance of Analytical Marketing and Data Mining: Towards an ERM Structural Framework

The Customer Relationship Management System consists of several components (Fig. 1): the CRM system aims to drive and animate the relationship with the customer through actions launched on the different channels of contact with the customer. The complexity of the system comes from the heterogeneity of the different components, the large number of customers and the offers that can be made to them, and finally the difficulty of animating a multi-channel relationship that must bring to each customer the right one, offer at the right time on the right channel [15].

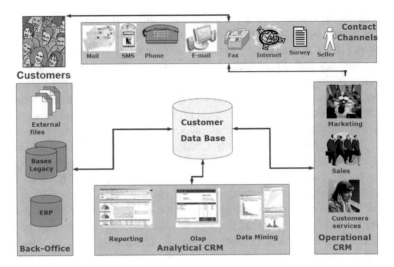

Fig. 1. CRM system (Source: adapted from [15])

The customer database contains all the data available on the customers (and sometimes also the prospects). It often has the structure of a data warehouse, whose data model is optimized to produce the reports. The concepts of the data warehouse were introduced in the 90's by [16, 17] and have since been widely developed to give birth to Business Intelligence (BI or business intelligence). The customer database is powered by multiple sources. CRM Analytics uses the data stored in the customer database. It has three main functions implemented by Business Intelligence techniques (reporting & OLAP) or data mining. Data mining analyzes are used either to enrich reports (predictive business intelligence) or to develop customer knowledge and build models for operational CRM actions that allow to launch targeted actions according to identified customer segments.

Therefore, based on this CRM system and transposing it to the political sphere, the implementation of activities that are part of the ERM practice involves:

(1) That information on citizen-electors, which is distributed by different data sources, be stored in an appropriate repository, thereby preserving a unique vision of the citizen-elector,

(2) That the information be analyzed using the appropriate data analysis tools, in order to obtain information about the citizen-elector or a group of citizen-electors,

(3) That a set of appropriate actions is automatically performed on the citizen-elector or a group of citizen- electors when a situation, event or specific behavior is detected.

The structural options related to the data repository and data analysis tools indicate that the proposed ERM system will be implemented using concepts and technologies associated with Business Intelligence systems. These systems combine data collection, data storage and knowledge management with analytical tools to present comprehensive and useful information for decision-making [18, 19].

3.2   Proposal for an ERM System Based on a Business Intelligence Infrastructure: A Generic View

The architecture of the ERM system is based on the technological infrastructure that traditionally supports Business Intelligence. A generic view of the ERM system is shown in Fig. 1 and described below. In order to integrate, store and keep all citizen-elector information from separate sources in a suitable data repository, it is proposed to store it in a data warehouse [20, 21] of which the structure must be precisely defined to support the ERM practice.

Subsequent data analysis, using appropriate tools (such as OLAP, data mining, querying, reporting) [22, 23] provides insights into citizen-elector. On the basis of this knowledge, a set of automatic actions will take place covering a citizen-elector or a group of citizen-electors, according to the ERM practices and activities defined by the political party.

The system must also verify the data collected over time and, where appropriate, alert messages are generated. For example, if a citizen-elector abstains from voting, the political party receives an automatic message. Subsequently, the impact of all actions on citizen-electors must be assessed.

The ERM system architecture consists of four main components: the data acquisition and storage component; the data analysis component; the interaction component; the evaluation component. An overview of this architecture is presented in Fig. 2.

The data acquisition and storage component is responsible for storing the citizen-elector data in a data warehouse, the structure of which will be modeled for this purpose. The data analysis component is responsible for obtaining information on citizen-electors. Stored data is analyzed using appropriate data analysis tools that identify trends and behaviors. The knowledge acquired is stored in a suitable data repository (knowledge base). Data analysis tools should allow: statistical analysis, querying, reporting, visualization of data, identification of citizen-elector profiles, behavior models.

The Interaction component is responsible for maintaining an adequate and effective relationship with the citizen-elector, through the knowledge acquired about them. The system must allow the definition and automatic execution of adequate actions on the citizen-elector or a group of citizen-electors. These are achievable with the help of technological tools allowing the interaction and the communication between the various participants of the system, as well as the automatic execution of personalized actions.

The Evaluation component is responsible for evaluating all the actions carried out and their impact, by monitoring the electoral behavior of the citizen-electors, through the analysis of different rates (abstention, neutrality, participation, among others) after the execution of the actions defined in ERM practice. The evaluation of the impact of the actions undertaken must also allow the redefinition of the actions themselves (and the ERM practice) if deemed necessary and/or adequate. The technological tools used in this component should also allow, among other things, the monitoring of different indicators, statistical analyzes and the execution of reports.

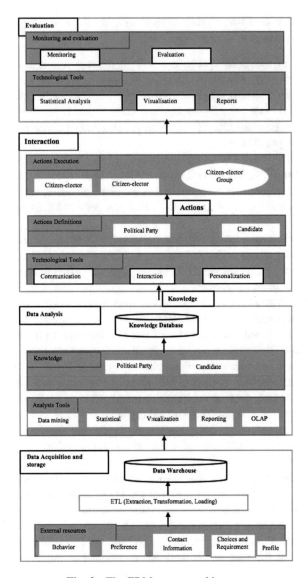

Fig. 2. The ERM system architecture

4   Conclusion

Through this article we proposed an ERM model that took place in different stages. The first step led us to the definition of the main components used in the concept, the practices and the ERM system. The second step was associated with the definition of the ERM system architecture and its main functionalities.

Moreover, to take into account the current political context in Morocco, where the voter abstention rate is high, it is necessary for political parties to adopt measures and procedures to reverse this trend. The context in which this change is to occur is characterized by strong competition between parties. We also expect political party members to play an active role in the process of managing the citizen-voter relationship. We also believe that the adoption of an ERM practice based on the proposed model, supported by a digital resource management system, is the key response to the change.

However, some limits are worth noting. Indeed, there is a lack of literature on the Elector Relationship Management and its system. Thing that pushed us to transpose and adapt the principles of the private sector to the political sphere in order to propose an ERM system. Note that this system is only a prototype and is still not implemented. But once implemented, it will be validated by the execution of a set of demonstration cases in different political parties in Morocco. We also think that the proposed concepts and practices can be validated by conducting a field survey.

References

1. Klaukka, G., van der Staak, S., Valladares, J.: The changing nature of political parties and representation. In: International Institute for Democracy and Electoral Assistance (International IDEA), The Global State of Democracy: Exploring Democracy's Resilience, pp. 98–118 (2017)
2. Bréchon, P.: La signification de l'abstention électorale, Texte rédigé à l'occasion d'un séminaire doctoral à l'Université libre de Bruxelles (ULB) (2010)
3. Adams, J., Dow, J., Merrill, S.: The political consequences of alienation-based and indifference-based voter abstention: applications to presidential elections. Polit. Behav. **28**(1), 65–86 (2006)
4. Muxel, A.: La vague de l'abstention, Science Po, Centre de recherche (2014)
5. Schellong, A., Langenberg, T.: Managing citizen relationships in disasters: Hurricane Wilma, 311 and Miami –Dade County. In: Proceedings of the 40th Hawai International Conference on System Sciences, Hawai, USA (2007)
6. Janssen, M., Wangenaar, R.: Customer relationship management in e-government: a dutch survey. In: European Conference on E-Government, 1–2 October, St Catherine (2005)
7. Zamanian, M., Khaji, M.R., Emamian, S.M.S.: The value chain of citizen relationship management (CzRM): a framework for improvement. Afr. J BM. **5**(22), 8909–8917 (2011)
8. Fayerman, M.: Customer Relationship Management, New Directions for Institutional Research, vol. 113, pp. 57–67 (2002)
9. Payne, A.: Handbook of CRM. Achieving Excellence in Customer Management. Elsevier-BH (Butterworth-Heinemann) (2006)
10. Mishra, A.: Customer relationship management: implementation process perspective. Acta Polytechnica Hungarica **6**, 83–99 (2009)
11. Hahnke, J.: The Critical Phase of the CRM Lifecycle. Without CRM Analytics, Your Customer Won't Even Know You're There (2001). www.hyperion.com
12. Wu, J.: Customer relationship management in practice: a case study of hi-tech company from China. In: International Conference on Service Systems and Service Management, 30 June–2 July 2008, pp. 1–6. IEEE Computer Society (2008)
13. Yu, J.: Customer Relationship Management in Practice: a Case Study of Hi-Tech from China. IEEE Computer Society (2008)

14. Bull, C.: Strategic issues in a customer relationship management (CRM) implementation. Bus. Process Manage. J. **9**(5), 592–602 (2003)
15. Fogelman, F.: CRM Analytique, l'apport du data mining. In: Bennani, Y., Viennet, E. (eds.) Apprentissage Artificiel & Fouille de Données, Volume: Revue des Nouvelles Technologies de l'Information, RNTI-A-2. Cépaduès, pp 15–34 (2007)
16. Inmon, W.H.: Building the Data Warehouse, 2nd edn. Wiley, New York (1996)
17. Kimball, R., Reeves, L., Thornthwaite, W.: The Data Warehouse Lifecycle Toolkit: Expert Methods for Designing, Developing, and Deploying Data Warehouses. Wiley (1998)
18. Negash, S., Gray, P.: Business intelligence. In: Ninth Americas Conference on Information Systems (2003)
19. Han, J., Kamber, M.: Data Mining: Concepts and Techniques. Morgan Kaufmann Publishers (2001)
20. Inmon, W.H.: Building the Data Warehouse. Wiley (2005)
21. Cunningham, C., Song, I.Y., Chen, P.P.: Business Intelligence: Data Warehouse Design to Support Customer Relationship Management Analyses, ACM - The Guide (2004)
22. Thomsen, E.: OLAP Solutions. Building Multidimensional Information Systems. Wiley (2002)
23. Berry, M.J.A., Linoff, G.S.: Data Mining Techniques for Marketing, Sales, and Customer Relationship Management. Wiley (2004)

Energy Management in WSNs

Yassine Rayri[✉], Hatim Kharraz Aroussi, and Abdelaziz Mouloudi

University Ibn Tofail, Kénitra, Morocco
rayri.yassine@gmail.com

Abstract. The new technology had created "Wireless sensor". This micro device allows measuring a physical quantity from the environment and transforms it autonomously into a digital value that can be processing. The deployment of several wireless sensors communicating by wireless radio, form a Wireless Sensor Network (WSN). This system can collaborate and collect information from its environment and send them to a Base Station (BS). The collection, processing and transmission of data are the main factors of the dissipation of the energy for the wireless sensors, since these Battery-powered sensors is limited in energy, and it is usually impossible to recharge or replace it, knowing that sensors are generally distributed in places which are difficult to reach. Indeed, the lifetime of the network is one of the major constraints facing in WSN. Therefore the energy consumption of the sensors plays an important role in the network lifetime. Among scientific research developed to improve the lifetime of wireless sensors network is the integration of a new techniques of routing protocols existing. In this paper, we propose a new routing protocol called EDE protocol (Equitable Distribution Energy) based on clustering. The purpose from our protocol is to guarantee an equal distribution of the energy across the network and eliminate the energy holes (black area), without reducing the service life of networks. The Comparisons will be made and studied in order to validate the evaluation of the performances of the proposed technique.

Keywords: Wireless sensor · Clustering · Leach · Equitable distribution energy

1 Introduction

The wireless sensor is a micro-device that measures a physical quantity, such as temperature, light, or the movement from the environment and transforms it autonomously into a digital value that can be processed, and transferred to the base station.

Deployment of multiple wireless sensors communicating over wireless radio forms a wireless sensor network. Several constraints prevent the proper deployment of these networks. Energy consumption is one of the most important challenges for wireless sensor networks. However, energy saving is one of the major issues in these networks, and it is difficult, if not impossible, to replace the sensors or their battery because of the deployment location that is often inaccessible, especially that the purpose of traditional application scenarios is to deploy nodes in an unattended domain for months or years [1–3]. The lifetime of a sensor network is the period during which the network may, as appropriate, maintain sufficient connectivity, cover the entire domain or maintain the rate of loss of information beyond a certain level [4].

M. Ezziyyani (Ed.): AI2SD 2019, LNNS 92, pp. 127–136, 2020.
https://doi.org/10.1007/978-3-030-33103-0_13

The equitable distribution of energy throughout the network is another trouble, especially when the station base is not in the middle of the network [5]. In this work, we propose an improvement of routing protocols (EDE: Equitable Distribution Energy) based on Leach protocol by reducing the overload of CH (cluster-head), in order to properly distribute the energy overall network; avoid the energy holes (death node). This EDE protocol consists on creating a new node (Transfer-node) in step up phase, within the cluster that is only responsible of transferring data into the base station (BS) in steady up phase. Our network will be structure as a producer-consumer pattern in order to organize intra-cluster exchanges. Also the protocol will maintain more or less the same energy performance in terms of lifetime of the network. At last, we will demonstrate the effectiveness of our proposed protocol by comparing it to other well-chosen protocols especially LEACH and SEP protocol [6, 7]. The simulations will be executed in Matlab.

The rest of the paper is organized as follows: In Sect. 2, related works. In Sect. 3, we are going to expose the problem of energy distribution in WSNs. In Sect. 4, we will define our proposition, protocol EDE. Simulation results and discussion are presented in Sect. 5, and finally, a conclusion and perspectives are given in the last section.

2   Related Works

Energy consumption remains a challenge for the research community, because of its paramount importance to ensure good performance of the network [8].

Maximizing the life of the network generally aims to minimize the energy consumption of the sensors. But also we can optimize the life of a network without necessarily maximize it, indeed, to ensure proper management of energy consumption of the system as a whole or to ensure an equitable distribution of energy throughout the network to maintain full coverage and a smooth operation of the system.

Mismanagement of energy consumption in the networks can produce so-called energy holes. We have two types of this problem, the first, near to the base station for multi-hop transmission topology; the second is far from the base station for the direct transmission topology. To solve this problem of energy holes or black zone, several approaches are proposed.

Sensors located closer to the base station, especially for multi-hop relays [9], participate in more data transfers (more load); sensors situated in this region should have a higher density.

The method has been proposed to use mobile sensors [10], allowing sensors to move to meet density requirements, In order to extend the life of the network and achieve a balanced energy consumption. But this approach has also shortcomings like, the investment will be more expensive for network deployment and also the process may take longer to collect all the data, which is not desirable for real-time monitoring [11].

A pixel-based transmission mechanism is adopted by [12], to reduce the duplication of the same messages, in order to obtain equitable energy consumption per node. The default of this proposal is that each node must know its location [13].

Among the solutions proposed for this problem, we have the [14] approach of dividing the coverage area of the network into systematic cells and layers. These cells are classified into primary and secondary cells. The problem with this approach is the difficulty of deploying network node sensors.

Clustering is one of the techniques created to maximize life and balance energy in wireless sensor networks. The Leach protocol [15] is classified among one of the reference protocols based on clustering [16]. This protocol has really succeeded in minimizing energy dissipation in the network, but it has not achieved the objective of solving the problem of energy holes or black areas (details in Sect. 3).

In the rest of this paper, we are focused on the second type black area for a far to the base station.

3 Problem of Energy Distribution

Typically, the cluster-head is the most active node of the cluster, because it manages all nodes member in the cluster. In effect, after the election of the cluster-head, it is in charge of the creation and the organization of the cluster, and then the data collection from member nodes, the treatment and the compression of this information in digital form, in order to transmit them to the Base Station [10]. In carrying out all these stains, the cluster-head loses a lot of energy, consequently it results a quick death of the cluster head in the network.

On another side, the location of the base station also creates a problem of energy. Indeed, cluster-head which are located away from the BS consume more energy, because the transmission distance to the BS is significant, knowing that transmission energy [6, 17] is:

$$E_T(k, d) = \begin{cases} E_{elec} * k + \in_{fs} * k * d^2 & d < d_0 \\ E_{elec} * k + \in_{amp} * k * d^4 & d \geq d_0 \end{cases} \qquad (1)$$

Where k is the message length in bits, E_{elec} is the data rate of each sensor node, d is the distance between the node transmitter and receiver, $(\in_{fs}, \in_{amp})$ is an amplifier energy parameters and d_0 is the threshold distance between multi-path fading model and the free-space model.

In addition, the areas that make up these sensors die completely. Therefore, when we work in a large area especially when we have a large number of nodes, sensors far from BS in the network lose a lot of energy [8] compared to other sensors. In this case, we can say that we do not have an equitable distribution of energy across the network. Reflecting what goes on the Leach protocol. Leach is based on the creation of clusters and the election of cluster-heads. CHs transmit data collected from member nodes directly to the base station. Thus, for the LEACH protocol, there is the problem of the overload of the CH that loses the energy of sensing, aggregation, reception and transmission.

We obtain two cases (Fig. 1):

First Case. There will always be a creation of CHs in the remote area of the BS to maintain a uniform distribution of CHs across the network. In this case, the CHs nodes will quickly lose energy.

Second Case. After an energy threshold, we will not have any CHs in the remote area of the BS. In this case, the member nodes will lose energy due to the transmission of the data collected each time to the distant CHs.

Hence, in both cases we get black areas.

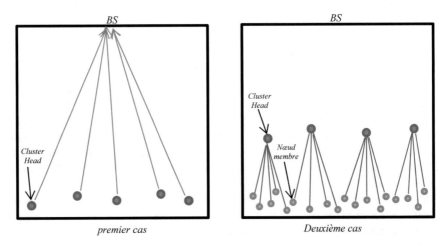

Fig. 1. (First case) energy dissipation of CH and (second case) energy dissipation of member nodes

4   EDE Protocol

4.1   Definition of the EDE Protocol

To reduce the functionalities of the cluster-head, we propose a new concept of routing (EDE protocol) based of Leach protocol, which consists in creating a node that is only responsible of the transfer of the data. The cluster-head delegates the task of the transmission of packets to the base station to another node of the cluster. This node called Transfer-Node (TN).

The EDE protocol process takes place on two phases same as the Leach protocol. In step up phase we add the selection of NT node. Indeed, in a defined cluster, the energy capacity of the chosen Node Transfer must be greater than or equal to the energy of the cluster-head. Otherwise; the latter will be TN node. Also, the difference in the steady up phase is that the CH transmits the data to the TN node.

We consider that the advanced EDE protocol has the same properties as the EDE protocol, but with normal nodes and advanced nodes that have more energy (heterogeneous node sensor network).

4.2   Impact Energy

We can expose cluster energy E_{C-CC} for a classic clustering routing protocol as follows:

$$E_{C-CC} = E_{S-all} + E_{P(CH)} + E_{R(CH)-all} + E_{T(CH)-TN} + E_{T(CH)-S} \qquad (2)$$

And the energy of a cluster for our proposition as follows:

$$E_{C-EDE} = E_{S-all} + E_{P(CH)} + E_{R(CH)-all} + E_{R(TN)-CH} + E_{T(TN)-S} \qquad (3)$$

Where E_{S-all} is the sensing energy, $E_{P(CH)}$ is the processing energy from cluster-head, $E_{R(CH)-all}$ is the reception energy consumed by the CH by receiving data from the member nodes, $E_{T(CH)-TN}$ is the transmission energy from CH to the TN node, is $E_{T(CH)-S}$ the transmission energy from CH to the sink, $E_{R(TN)-CH}$ is the reception energy consumed by the TN node by receiving data from the member nodes, and $E_{T(TN)-S}$ is the transmission energy from TN node to the sink. We consider that,

$$E_{C-CC} = E_{C-EDE} \qquad (4)$$

and

$$E_{CH-CC} = E_{P(CH)} + E_{R(CH)-all} + E_{T(CH)-TN} + E_{T(CH)-S} \qquad (5)$$

and

$$E_{CH-EDE} = E_{P(CH)} + E_{R(CH)-all} \qquad (6)$$

According to the above

$$E_{CH-CC} = E_{CH-EDE} + E_{R(TN)-CH} + E_{T(TN)-S} \qquad (7)$$

We assume that E_{TN} is the energy consumed by TN node,

$$E_{TN} = E_{R(TN)-CH} + E_{T(TN)-S} \qquad (8)$$

$$E_{CH-CC} = E_{CH-EDE} + E_{TN} \qquad (9)$$

And finally, we have the energy dissipated by CH is greater than the energy by TN.

$$E_{CH-CC} \geq E_{CH-EDE} \qquad (10)$$

The remaining energy, since we have the same energetic mass, is shared on the other nodes that are selected TN nodes (Eq. 9).

5   Simulation Results and Discussion

5.1   Simulation Environment

We chose to compare with the Leach protocol, since our proposed protocol is an improvement of the Leach protocol. In order to obtain a diversity of topology on the networks, we chose to compare with the SEP protocol because it is based on heterogeneous sensor networks [7]. In SEP protocol m is the fraction of advanced nodes and α is the additional energy factor between advanced and normal nodes.

The data exchanges between the member nodes, CH node and TN node are done according to a producer-consumer pattern (Fig. 2), to synchronize data sharing processes.

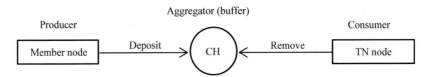

Fig. 2. Producer-consumer pattern applied on the EDE protocol

The wireless sensors deployed on an area of 100 m * 100 m with a base of station located at the point (50, 100). We use Matlab environment to run simulation (Table 1).

Table 1. Parameters table

Parameter	Value
Initial energy of nodes	0.25 J
Transmitter and receiver energy	50nj/bit
Aggregation energy	5nj/bit
Data packet length	4000 bit
Amplifier energy E_{fs}	10pj/bit/m2
Amplifier energy E_{amp}	0.0003pj/bit/m4

5.2   Simulation Results

In this simulation, we compare our protocol EDE with Leach and SEP protocol on energy distribution in the network and energy conservation efficiency. We consider that all nodes have an equal initial energies and each death node excluded from the next round.

First Simulation (Energy distribution). In Fig. 3(a), where treated by the classic Leach protocol, clearly shows that the lower part of the network represents a black area in which all the sensors are dead. Thus, we cannot collect information from this area that is not covered, which leads to reduced network performance.

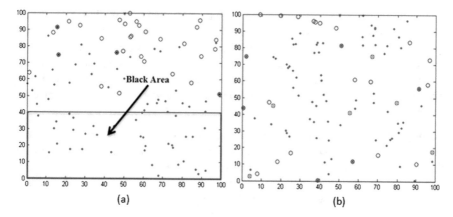

(a)   (b)

Fig. 3. Alive nodes (circles) and dead nodes (dots) for (a) Leach protocol where blue circles represent the member nodes, the black circles represent the cluster heads and the red dots represent the dead nodes, and for (b) EDE protocol where blue circles represent the member nodes, the red circles represent the cluster head, the green circles represent the transfer nodes and the red dots represent the dead nodes.

Figure 3(b) illustrates, treated by EDE in which we show the death of the nodes well distributed across the network. Indeed, we note that there are nodes located in the lower part of the network, even on the abscissa axis. Consequently, we can have a data collection on the whole area of the network.

Now, we compare our EDE protocol with the SEP protocol with $\alpha = 0.2$ and m = 1.

Note also that, for SEP, Fig. 4 (a) shows that the nodes of the lower part of the network are dead, which means that distribution of energy is not fair on the network, contrary to our proposition (Fig. 4) (b) which we have a good distribution of energy.

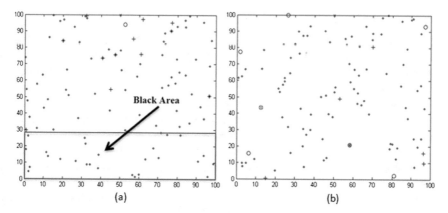

(a)   (b)

Fig. 4. Alive nodes (circles, plus sign) and dead nodes (dots) for (a) SEP protocol where blue circles represent the member nodes, the black circles represent the cluster heads, the red dots represent the dead nodes, and nodes in the form of the "plus sign" represent the advanced nodes, and for (b) EDE advanced protocol where blue circles represent the member nodes, the red circles represent the cluster head, the green circles represent the transfer nodes, the red dots represent the dead nodes, and nodes in the form of the "plus sign" represent the advanced nodes.

In addition, for the SEP protocol, the normal nodes died first, while for the proposed protocol, as shown in Fig. 4 (b), there remain the two categories of nodes, normal and advanced. This implies that we do not have an overload on some nodes.

From the simulations above illustrate by Figs. 3 and 4, we conclude that the EDE protocol based on creating the TN nodes, has guaranteed a fair distribution of energy better than the Leach and SEP protocol. That is, the network life of the coverage point of view is improved.

Second Simulation (Energy Consumption). In this simulation, we compare energy conservation efficiency between EDE and leach protocol, also EDE and SEP protocol. We consider that all nodes have an equal initial energy.

In Fig. 5, we have focused on the case of the network that consists of 100 nodes. We are comparing the number of round with the dead node.

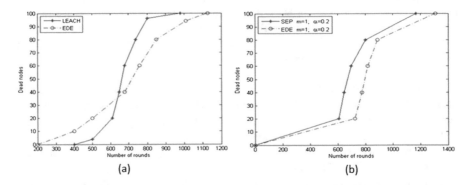

Fig. 5. Number of rounds in relation with dead nodes for (a) between EDE and leach protocol, and (b) between EDE and SEP protocol

Although, EDE protocol starts to lose its own nodes, from round number 200, but the Leach protocol begins to lose its nodes from round number 402, as shown in Fig. 5 (a). In round 648 the Leach protocol loses energy faster than the EDE protocol. The total number of rounds of the EDE protocol is slightly higher than the number of rounds of the Leach protocol.

Next, we are comparing the energy conservation between the SEP protocol and the EDE protocol with 100 nodes. We consider that we have 100 nodes distributed in normal and advanced with $\alpha = 0.2$ and m = 1.

The Fig. 5(b) also shows that the rounds number of the EDE protocol which equal 1311 is slightly higher than the number of rounds of the SEP protocol which equals 1189.

According to Fig. 5, we can conclude that the EDE protocol maintains the longevity performance of the network lifetime compared to the Leach and SEP protocols.

6 Conclusion and Perspective

Several areas of research that have worked on optimizing the energy dissipation of wireless sensor networks have led to the creation of multiple routing protocols; each of them has advantages and disadvantages according to the criteria of the topology of the network and the sensors. Scientific research in the energy field is focused, generally on maximizing the life of wireless sensors. However, it must be remembered that the good distribution of energy throughout the network has an impact on the lifetime of the entire network, sometimes knowing that the loss of a part of the network makes the whole network defective network and that causes the problem of energy holes or black area. The integration of the transfer node (Transfer-Node) and the producer-consumer model into a hierarchical protocol has given birth to a contribution which is the EDE (Equitable Distribution Energy) protocol. The latter has shown an efficiency that allows obtaining a fair and equitable distribution of energy on the entire network, which solved the problem of energy holes. Indeed, the results of the simulation verify the effectiveness of ours analyzes and proposed solutions.

The work done in this paper also reveals perspectives that deserve continuity in research. Thus, in the future work we will improve our EDE protocol to have better results in energy conservation and also, test this proposed approach on other hierarchical routing protocols.

References

1. Clarea, L.P., Pottieb, G.J., Agrea, J.R.: Self-organizing distributed sensor networks. In: Proceedings SPIE 3713, Unattended Ground Sensor Technologies and Applications, pp. 229–237 (1999). https://doi.org/10.1117/12.357138
2. Fu, C., Jiang, Z., Wei, W., Wei, A.: An energy balanced algorithm of LEACH protocol in WSN. IJCSI Int. J. Comput. Sci. Iss. 10(1), 354–359 (2013)
3. Anastasi, G., Conti, M., Di Francesco, M., Passarella, A.: Energy conservation in wireless sensor networks: a survey. Ad Hoc Netw. 7(3), 537–568 (2009). https://doi.org/10.1016/j.adhoc.2008.06.003
4. Kacimi, R.: Techniques de conservation d'énergie pour les réseaux de capteurs sans fil. Doctorat de l'universite de toulouse (2009)
5. Pawar, K., Kelkar, Y.: A survey of hierarchical routing protocols in wireless sensor network. Int. J. Eng. Innovative Technol. 1(5), 50–54 (2012)
6. Heinzelman, W.B., Chandrakasan, A., Balakrishanan, H.: An application-specific protocol architecture for wireless micro sensor networks. IEEE Trans. Wireless Commun. 1(4), 660–670 (2002). https://doi.org/10.1109/twc.2002.804190
7. Smaragdakis, G., Matta, I., Bestavros, A.: A stable election protocol for clustered heterogeneous wireless sensor networks. In: Proceedings of the 2nd International Workshop on SANPA 2004, Massachusetts, U.S, p. 1 (2004)
8. Dhawan, H., Waraich, S.: A comparative study on LEACH routing protocol and its variants in wireless sensor networks: a survey. Int. J. Comput. Appl. 95(8), 21–27 (2014). https://doi.org/10.5120/16614-6454

9. Chen, Z., Chen, K.: An improved multi-hop routing protocol for large-scale wireless sensor network based on merging adjacent clusters. J. Softw. **8**(8), 2080–2086 (2013). https://doi.org/10.4304/jsw.8.8.2080-2086

10. Cardei, M., Yang, Y., Wu, J.: Non-uniform sensor deployment in mobile wireless sensor networks. In: IEEE International Symposium on a World of Wireless, Mobile and Multimedia Networks (WoWMoM 2008), pp. 1–9 (2008). https://doi.org/10.1109/wowmom.2008.4594823

11. Li, J., Mohapatra, P.: Analytical modeling and mitigation techniques for the energy hole problem in sensor networks. Pervasive Mob. Comput. **3**(3), 233–254 (2007). https://doi.org/10.1016/j.pmcj.2006.11.001

12. Jia, J., Chen, J., Wang, X., Zhao, L.: Energy-balanced density control to avoid energy hole for wireless sensor networks. Int. J. Distrib. Sens. Netw. **2012**(812013), 1–10 (2012). https://doi.org/10.1109/access.2017.2728367

13. Sharma, R.: Energy holes avoiding techniques in sensor networks: a survey. Int. J. Eng. Trends Technol. (IJETT) **20**(4), 204–208 (2015). https://doi.org/10.14445/22315381/ijett-v20p239

14. Subir, H., Amrita, G., Sanjib, S., Avishek, D., Sipra, D.: A lifetime enhancing node deployment strategy in WSN. In: Lee, Y.-H., Kim, T.-H., Fang, W.-C., Ślęzak, D. (eds.) FGIT 2009. LNCS, vol. 5899, pp. 295–307. Springer, Heidelberg (2009). https://doi.org/10.1007/978-3-642-10509-8_33

15. Heinzelman, W.R., Chandrakasan, A., Balakrishnan, H.: Energy-efficient communication protocols for wireless microsensor networks. In: Proceedings 33rd Hawaii International Conference on System Sciences, January 2000, vol. 2, pp. 1–10 (2000). https://doi.org/10.1109/HICSS.2000.926982

16. Liu, X.-X.: A survey on clustering routing protocols in wireless sensor networks. Sensors **12**(8), 11113–11153 (2012). https://doi.org/10.3390/s120811113

17. Liao, Q., Zhu, H.: An energy balanced clustering algorithm based on LEACH protocol. In: Proceedings of the 2nd International Conference on Systems Engineering and Modeling (ICSEM-13), pp. 72–77 (2013). https://doi.org/10.2991/icsem.2013.15

Estimation of the Best Measuring Time for the Environmental Parameters of a Low-Cost Meteorology Monitoring System

Laura García[1,2], Lorena Parra[1], Jose M. Jimenez[1,2], Jaime Lloret[1(✉)], and Pascal Lorenz[2]

[1] Integrated Management Coastal Research Institute,
Universitat Politècnica de València, Valencia, Spain
laugarg2@teleco.upv.es, loparbo@doctor.upv.es,
{jojiher, jlloret}@dcom.upv.es
[2] Network and Telecommunication Research Group,
University of Haute Alsace, Colmar, France
lorenz@ieee.org

Abstract. Meteorology monitoring is crucial for implementing smart agriculture systems. These systems should employ as few power as possible in order to avoid workers going to the fields to replace batteries. Thus, the collection and forwarding of data should be reduced as much as possible. However, the time interval to be employed should be large enough for the data to be accurate and so as to avoid data loss. In this paper, we examine different time intervals for data acquisition utilizing our proposed algorithm for our low-cost meteorology monitoring system. Real data has been analyzed with time intervals from 5 to 60 min. Results that the best time interval is 25 min for temperature, 45 min for humidity and 5 min for light.

Keywords: Precision agriculture · ESP32 · Algorithm · Low-cost · Decision making

1 Introduction

The predictions related to the world population, presented by different international organizations, suggest that there will be a large and constant increase in population in the coming years. Organizations such as the United Nations [1] predict that the world population will reach 8.5 billion people by 2030, and 9.7 billion by 2050. This increase in population brings with it the need for a very important increase in food production. But given this need, there are great difficulties to increase current production, among other reasons due to the important changes that are taking place over the last few years in meteorology, water resources are becoming increasingly scarce every day.

Due to the causes previously exposed, it is necessary that agriculture advances in the future and becomes an intelligent agriculture. Smart agriculture is achieved by applying Information and Communication Technologies to traditional agriculture [2]. Thanks to their employment, farmers and engineers can use smart tools [3], through which they can be more efficient and consume smaller amounts of natural resources.

© Springer Nature Switzerland AG 2020
M. Ezziyyani (Ed.): AI2SD 2019, LNNS 92, pp. 137–144, 2020.
https://doi.org/10.1007/978-3-030-33103-0_14

Implementing this type of solutions can optimize the moments to temporize vital actions in the harvests, like the irrigation, sowing or the moment to add fertilizers to the terrain. In addition, through Artificial Intelligence, patterns can be determined, which can be used to generate intelligent systems that are very useful when making decisions.

In smart agriculture, it's vital to have information about multiple parameters that allow us to establish precision agriculture [4]. Among others, we can highlight as fundamental the obtaining of information in real time about temperature, humidity, precipitation or light.

Our proposal presents the evaluation of different time intervals to be applied to monitor meteorological parameters while reducing the amount of data gathered and transmitted by the system while maintaining the necessary information for decision making. We also present the results of real measurements obtained in an area of Mediterranean climate. This work is part of the implementation of our proposed smart irrigation systems [5].

The rest of the paper is organized as follows. Section 2 presents the related work. The system description is depicted in Sect. 3. The results are discussed in Sect. 4. Finally, the conclusion and future work is presented in Sect. 5.

2 Related Work

The Related work on low-cost meteorological monitoring systems and energy-saving algorithms is going to be presented in this section.

Due to the importance of meteorology monitoring, researchers have developed several meteorology monitoring systems. F. J. Mesas Carrascosa et al. presented in [6] an environmental monitoring system for precision agriculture. The system is comprised of the Arduino ATMega2560 microcontroller, analogue and digital temperature and humidity sensors, a light sensor that detects the presence and absence of light, a clock, a Bluetooth module, an SD card module and a photovoltaic panel. The system was able to forward the data to a database for further processing. Real measurements were performed in a farm located in the Spanish city of Cordoba. The price of their proposal is below 100€ and results were similar to those of a real meteorological station. Furthermore, a wireless weather station that included a web interface for users was developed by Hardeep Saini et al. in [7]. The system was able to monitor temperature, humidity, pressure, wind direction and wind speed. Zigbee was employed for data transmission and alerts were generated according to the measured parameters. Results showed the accuracy of the proposed system.

These systems need to be deployed outdoors, thus the use of energy-efficient algorithms is necessary. Thar Baker et al. presented in [8] an energy-aware algorithm for IoT applications based on cloud called (E2C2). The algorithm allows utilizing the least possible number of services needed to complete the requirement. The proposed algorithm was compared to four other existing proposals. Results showed a better performance of the proposed algorithm regarding energy efficiency and least number of services used to complete a task in an optimal manner. Moreover, Ali Hassan Sodhro et al. developed in [9] an energy-efficient algorithm for e-health electrocardiogram wearable devices. The proposed algorithm called ETPC is then compared to other transmission power control algorithms. Results showed energy savings of 35.5% compared to those of other TCP solutions. However, the resulting packet loss was slightly higher than other TCP algorithms.

In this paper, an energy saving algorithm is going to be applied for our developed meteorology monitoring system. As opposed to other existing systems, our meteorology monitoring system is able to determine the amount of light measured in lux as well as temperature, humidity and the presence of rain. The algorithm has been applied to real measures obtained from a Mediterranean climate.

3   System Description

In this section, the description of the meteorological monitoring system for agriculture is going to be presented.

There are several environmental parameters whose knowledge is important to farmers and the development of a comprehensive system for precision agriculture. Temperature is one of the most important factors that affect plants. The correct growth of plants, blooming, photosynthesis or water and nutrient absorption are some of the aspects that depend on the temperature of the environment. To measure temperature, on this prototype the DHT11 sensor has been utilized. This sensor has been utilized to measure humidity as well. Humidity affects plant transpiration, its photosynthesis and its growth. Low levels of humidity are related to withered plants, small sized leaves or burnt or dried leave tips. High humidity levels are related to higher root and leave diseases, nutrient deficiencies or weak growth. The rain determines whether the irrigation should be activated or not. Furthermore, high amounts of rain should be accompanied of the implementation of proper drainage so as to avoid the excess of humidity in the soil. A rain sensor was utilized so as to determine the presence or absence of rain. Lastly, the amount of light is related to the photosynthesis process of the plants and may affect the constitution of the fruit.

Apart from the sensors described previously, the system is comprised of an ESP32 Devkit V1 microcontroller [10] Furthermore, an SD card is incorporated to the prototype so as to store the information in case of the malfunction of the wireless connection or the server and the power is provided through a 12 V solar panel and a set of batteries. The system is presented in Fig. 1. The DHT11 sensor is located inside a protective plastic with open spaces to let the air flow. The open spaces are protected with a net to avoid the incursion of insects. The other sensors are located on the bigger encapsulation. The solar panel is supported by an aluminum structure and the rest of the circuit and the microcontroller is located inside the encapsulation.

The algorithm of the system is presented in Fig. 2. Firstly, the thresholds of the parameters are set. Then the initial values are measured and used as reference values. The counter and variables of the different metrics are initialized afterwards. The system starts taking measures and calculates the averages of temperature and humidity and the maximum value for luminosity. Then, when the counter reaches the dt time, if there have been changes, the data is forwarded to the database and the reference values are reset. If there have not been any changes, the previous values are maintained.

4 Results

The results of the simulation of applying our previously proposed algorithm to the meteorological monitoring system are presented in this section. To determine the best *dt* for each parameter, the followed approach has been to evaluate the accumulate relative error (ARE) of the data in the database (DB) for 15 h. The data is real environmental data gathered by our environmental monitoring system presented in the previous section. This 15-hour period includes two time lapses with big changes at the beginning and at the end of the experiment (sunset and sunrise), and a long time lapse with minor changes (the night). We consider that the best *dt* is the highest one that allows to have an ARE lower than 5%. Therefore, we minimize the number of packets and keep a good data accuracy in the DB.

First, the data about temperature when different *dt* are utilized in the algorithm is presented. The values of temperature in de DB with different dt are presented in Fig. 3. We can see that, at the beginning, of the data set the data values of temperature are decreasing slowly due to the sunset. During the night the temperature still decreases more leisurely. At the end of the experiment, after the sunrise, the temperature starts to increase. When different dt are used, the values of temperature in the DB change, and some fluctuations are lost. For example, for the *dt* of 50, 55 and 60 min, the minimum temperature (7 °C) are not recorded in the DB, and the increment of temperature during the sunrise is not correctly gathered with *dt* higher than 25 min.

Regarding the ARE in temperature values of the DB, see Fig. 4, we can affirm that, ARE increases dramatically when the *dt* is higher than 25 min. ARE for *dt* between 30 and 55 min are almost the same and increases again with *dt* of 60 min. Therefore, considering the ARE during this experiment and the aforementioned information in Fig. 4, the best *dt* for temperature sensing, according to our data, is 25 min.

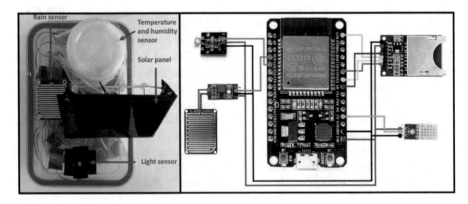

Fig. 1. Connection of the elements of the system.

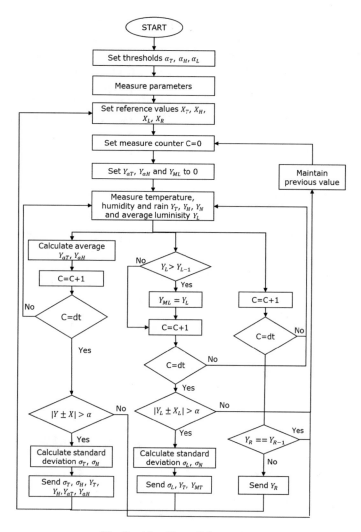

Fig. 2. Algorithm of the system.

Following, the data stored in the DB that is gathered by the Humidity sensor is shown, see Fig. 5. During the sunset and the first hours of the night, the humidity increases. Then, after the sunrise, the humidity decreases due to the increase of temperature. As the changes in humidity are slower than in the case of temperature, no big differences can be seen in the values of the DB with the different dt beyond a delay on the notification of the changes. The maximum and minimum values are gathered with all the dt and a good representation of data variability is archived in all the cases.

In terms of ARE of humidity data, see Fig. 6, it increases almost constantly as the dt upsurges. The only dt that overtakes the limit of 5% is the $dt = 60$ min. Nonetheless, the dt of 50 and 55 min are close to reaching 5%. Therefore, we can say that to ensure that ARE do not exceeds the 5%, it is better to use a dt of 45 min for the data of humidity.

Finally, we focus on the data of illumination, see Fig. 7. This data presents the highest changing rate in sunrise and sunset. Consequently, we expect to have high mismatches in the data when high *dt* are used. In this case, the dt higher than 5 (the original data) are not able to record the maximum light value. Moreover, the dt higher than 15 min cannot gather properly the abrupt changes in the sunrise and the sunset. Thus, according to the capability to record the maximum, minimum values and the variations in the data, the maximum acceptable *dt* is 15 min, as seen in Fig. 8.

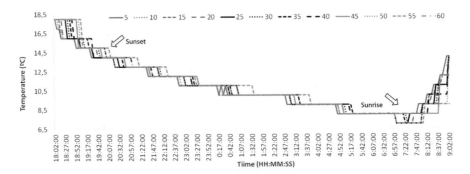

Fig. 3. Value of Temperature in the DB with different *dt* in algorithm X.

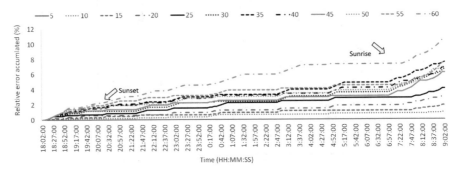

Fig. 4. ARE of Temperature in the DB with different *dt* in algorithm X.

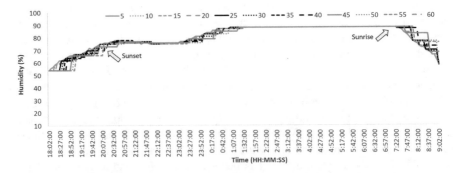

Fig. 5. Value of Humidity in the DB with different *dt* in algorithm X.

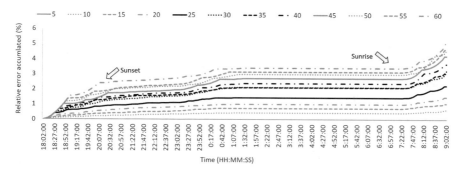

Fig. 6. ARE of Humidity in the DB with different *dt* in algorithm X.

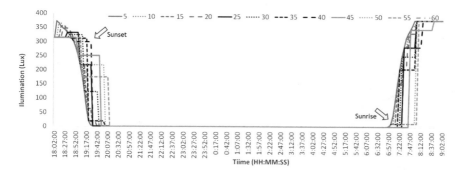

Fig. 7. Value of Illumination in the DB with different *dt* in algorithm X.

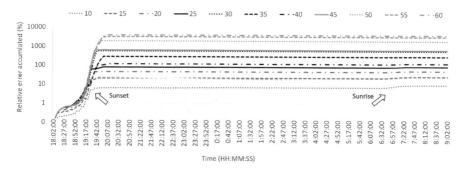

Fig. 8. ARE of Illumination in the DB with different *dt* in algorithm X.

As it can be seen, the obtained results vary from 45 min to 5 min depending on the parameter. Other environmental monitoring systems such as [6], perform a measure each hour. The selected time period must serve the required purpose. Our results indicate the best time period for each parameter in the Spanish Mediterranean weather in order to preserve the information of the changes in the weather. However, other climates may require different settings and other applications may not need much accuracy.

5   Conclusion and Future Work

In this paper, we have examined different time interval configurations for the algorithm of our presented meteorology monitoring system so as to determine the shortest time period that allows maintaining the required information. The results show that each parameter has a different dt that ensures a good representation of data in the DB: $dt = 25$ min for temperature, $dt = 45$ min for humidity and $dt = 5$ min for illumination. However, as the smallest dt forces to the system to send the data each 5 min, it will be necessary to evaluate the possibility of using a variable dt which can change according to the events. Having a shorter dt during sunrise and sunset, larger dt during the day and night and allow to the system to decide which dt should be used according to the data gathered and their changes.

For future work we are going to develop a low-cost soil monitoring device for the soil monitoring activities that are actually performed by costly commercial solutions. We will apply this algorithm to the soil monitoring system as well.

Acknowledgments. This work has been supported by European Union through the ERA-NETMED (Euromediterranean Cooperation through ERANET joint activities and beyond) project ERANETMED3-227 SMARTWATIR.

References

1. Population. United Nations. https://www.un.org/en/sections/issues-depth/population/index. html. Accessed 26 Apr 2019
2. Cambra, C., Sendra, S., Lloret, J., García, L. An IoT service-oriented system for agriculture monitoring. In: 2017 IEEE International Conference on Communication (ICC), Paris, France, 21–25 May 2017, pp. 1–6 (2017)
3. Rodrigues, J.J.P.C.: Emerging technologies for smart devices. Netw. Protoc. Algorithms **5**, 28–30 (2013)
4. Karim, L., Anpalagan, A., Nasser, N., Almhana, J.: Sensor-based M2M agriculture monitoring systems for developing countries: state and challenges. Netw. Protoc. Algorithms **5**, 68–86 (2013)
5. García, L., Parra, L., Jimenez, J.M., Lloret, J., Lorenz, P.: Practical design of a WSN to monitor the crop and its irrigation system. Netw. Protoc. Algorithms **10**, 35–52 (2018)
6. Mesas-Carrascosa, F.J., Verdú Santano, D., Meroño, J.E., Sánchez de la Orden, M., García-Ferrer, A.: Open source hardware to monitor environmental parameters in precision agriculture. Biosyst. Eng. **137**, 73–83 (2015)
7. Saini, H., Thakur, A., Ahuja, S., Sabharwal, N., Kumar, N.: Arduino based automatic wireless weather station with remote graphical application and alerts. In: 3rd International Conference on Signal Processing and Integrated Networks, Noida, India, 11–12 February 2016, pp. 605–609 (2016). https://doi.org/10.1109/SPIN.2016.7566768
8. Baker, T., Asim, M., Tawfik, H., Aldawsari, B., Buyya, R.: An energy-aware service composition algorithm for multiple cloud-based IoT applications. J. Netw. Comput. Appl. **89**, 96–108 (2017). https://doi.org/10.1016/j.jnca.2017.03.008
9. Sodhro, A.H., Sangaiah, A.K., Sodhro, G.H., Lohano, S., Pirbhulal, S.: An energy-efficient algorithm for wearable electrocardiogram signal processing in ubiquitous healthcare applications. Sensors **18**, 923 (2018). https://doi.org/10.3390/s18030923
10. DOIT ESP32 DevKit v1 specification. https://docs.zerynth.com/latest/official/board.zerynth. doit_esp32/docs/index.html

Genetic Algorithm for Shortest Path in Ad Hoc Networks

Hala Khankhour[1]([⊠]), Jâafar Abouchabaka[1]([⊠]),
and Otman Abdoun[2]([⊠])

[1] Faculty of Science, Department of Computer Science, IBN Tofail University,
Kénitra, Morocco
hkhankhour@uae.ac.ma, Abouchabaka.depinfo@gmail.com
[2] Department of Computer Science Polydisciplinary Faculty, Abdelmalek
Essaadi University, Larache, Morocco
Abdoun.otman@gmail.com

Abstract. The decentralized nature of ad hoc wireless networks makes them suitable for a variety of applications, where the central nodes cannot be invoked and can improve the scalability of large map networks, the topology of the ad hoc network may change rapidly and unexpectedly. Mobile Ad hoc (VANET) are used for communication between vehicles that helps vehicles to behave intelligently during vehicle collisions, accidents…one of the most problems confronted in this network, is finding the shortest path (SP) from the source to the destination of course within a short time. In this paper Genetic Algorithm is an excellent approach to solving complex problem in optimization with difficult constraints and network topologies, the developed genetic algorithm is compared with another algorithm which contains a topology database for evaluate the quality of our solution and between Dijkstra's algorithm. The results simulation affirmed the potential of the proposed genetic algorithm.

Keywords: Genetic algorithm · Population size · Shortest path · Ad hoc · Optimal routing · VANET

1 Introduction

Optimization is very important in many areas, such as operational research, artificial intelligence, biology, mathematics and computer science. A large number of problems can be defined and described as optimization problems. Generally, these problems belong to the class of NP-hard problems that haven't a solution optimal for the dataset. The problem of unicast routing with QoS is an NP-difficult optimization problem, it is very difficult to solve it by the traditional exact methods. especially for big map networks [1]. To solve this problem, studies converge on the use of approximate methods generally inspired by nature. Approximate optimization methods include heuristics and metaheuristics.

In 1975, genetic algorithms was invented by John Holland inspired by the Drawin's theory of evolution [2]. GA is a metaheuristic most chosen to solve complex problems of optimization in science and industry requiring high calculation times [3].

© Springer Nature Switzerland AG 2020
M. Ezziyyani (Ed.): AI2SD 2019, LNNS 92, pp. 145–154, 2020.
https://doi.org/10.1007/978-3-030-33103-0_15

The GA procedure is based on natural selection and genetics. GA collects the solution of a problem as a genetic string where each candidature solution called a chromosome and their genes (parameters) have been used for encoding the problem [4]. Each group of individuals (animals, plant species …), also called population, gives rise to a next generation through sexual reproduction. This generation cross the individuals between them to give offspring possessing the characters of both parents. In addition to this crossing, mutations of characters involved randomly in the next generation of the population. Then, this new population undergoes a selection, a metaphor for natural selection only the individuals best adapted to the environment survive. Finally, in turn, this population will give rise by the same process to a new population, which will be even more efficient in its environment [5], the rest of this paper is organized as follows, Sect. 2 present the procedure of genetic algorithm, in Sect. 3 proposed a brief description for genetic algorithm for Ad Hoc and the developed genetic algorithm in ad hoc network. The implementation details are presented in Sect. 4. In Sect. 5 are given the simulation results and discussed. The paper is concluding in Sect. 6.

2   Optimization in VANET Network

Ad hoc vehicle networks (VANETS) have attracted a lot of attention in recent years as they have been used to improve road safety. some study tries to map the features of the VANET to the characteristics of big data to ensure that the VANET issues can be assumed as a Big Data problem, which can be handled by big data techniques and tools [6]. Despite VANET's promising potentials, a major problem lies in the design of robust communication routing models for route optimization. With so many dynamic factors militating against effective routing in VANET, one pertinent research issue entails constructing an all-encompassing metric that can guarantee reliable routing in VANET [7], However, the main problem related to it is the potential high speed of moving vehicles [8], on the other hand, most of the work is done on networks with 50 nodes as maximum [8, 9].

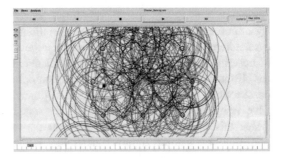

Fig. 1. Example of simulator network ns3

Figure 1 shows an example of network evaluate by the ns3 simulator, but it is very difficult to use it to solve complex problem. That is why we work by the methods of artificial intelligence among these methods are the genetic algorithms that will give good results we will explain them in detail in the chapters following.

Our challenge is to change the design of metaheuristics algorithm for solving the shortest path problem of a large-scale map of network, using genetic algorithm can significantly reduce the time also improve the quality of the provided solutions.

3 Genetic Algorithm

3.1 Presentation of the Genetic Algorithm

According to Lerman and Ngouenet (1995) a genetic algorithm is defined by [10]:

- chromosome/sequence: This is the set of genes of an individual (DNA chain).
- Population: a set of chromosomes or points in the search space.
- Environment: the research space.
- Fitness function: the positive function we seek to maximize.

3.2 The Course of a Genetic Algorithm Can be Divided into Five Parts

1. The creation of the initial population;
2. The evaluation of individuals;
3. The creation of new individuals;
4. The insertion of new individuals into the population;
5. Reiteration of the process.

3.3 The Steps of the Genetic Algorithm

- Choose the initial random population of individuals → Evaluate the fitness of individuals;
- Repeat;
- Randomly select the individuals to be used by the algorithm genetic;
- Generate new people using the crossover and the mutation;
- Evaluate the fitness of new individuals;
- Replace individuals of the population with new ones;
- Until certain criteria of stop. (number of iterations for example)

4 Genetic Algorithm for Ad Hoc

Ad hoc network is presented as a connected graph with N nodes. The chromosomes in the algorithm are sequences of integers and each gene represents a node ID that is selected randomly from the set of nodes connected with the node corresponding to its locus number. The gene of first locus is always taken for the source node and the last gene reserved for the destination node. the measured of optimization is the total cost of path between the source node and destination node, which is the goal, based on topological information database, but the length of the chromosome (route) it should not exceed the maximum of number the node in the network. The goal is to find the minimal cost path from source to destination [10].

4.1 Representation of a Population

The population represents a complete solution including routes, is a set of chromosomes, in the proposed algorithm, initially, a random population is generated, selects randomly genes (nodes) from the topological information database, the solution are string cannot achieve the destination, is rejected (unachievable) [10].

4.1.1 Evaluation of Fitness Function

In this paper every chromosome has its own quality evaluation function based on their fitness, the chromosome's quality function is determined by the chromosome's distance In population and the number of node, is defined as:

$$\text{fi} = \frac{1}{\text{li} - 1}$$

$$\sum C \text{ gi}(j), \text{ gi}(j + 1)$$

$$j = 1$$

Where fi represents the fitness value of the ith chromosome, li is the length of the ith chromosome, gi(j) represents the gene (node) of the jth locus in the ith chromosome, and C is the link cost between nodes In this paper, the selection of chromosomes make by the roulette system, is inspired by lottery wheels, so the best fitness value of chromosome has a great chance to get select into the next population.

4.2 Crossover

The crossings make it possible to simulate reproductions of individuals in the order to create new ones. It is quite possible to do random crosses [11]. In the proposed algorithm, the crossover operator chosen one-point crossover randomly, except the source and destination node, for each couple. chromosome can be cut in the middle of a gene. Chromosomes are of course usually much longer. The procedure is illustrated by Fig. 2.

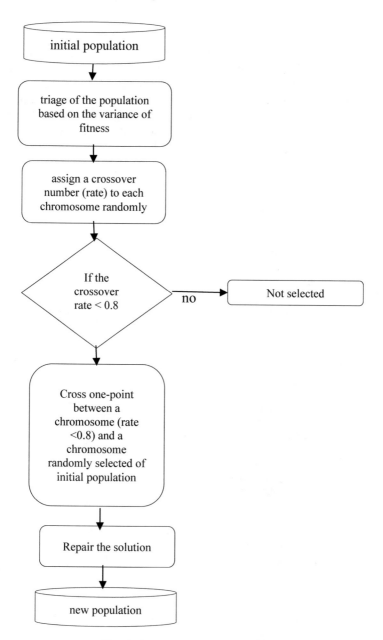

Fig. 2. The flow chart of crossover

4.3   Mutation

For each chromosome it has a small possibility to produce an illegal offspring after crossover. That is because it possibly introduces the central depot into the chromosome. The illegal offspring will be forced to mutate. A gene (node) is randomly selected first from the chosen chromosome ("mutation point"). After the mutation operator, the repair function finds the node already included twice in chromosome and it's should be deleted, and check if the next node and the previous node of the mutated node is valid [3]. The procedure is illustrated by Fig. 3.

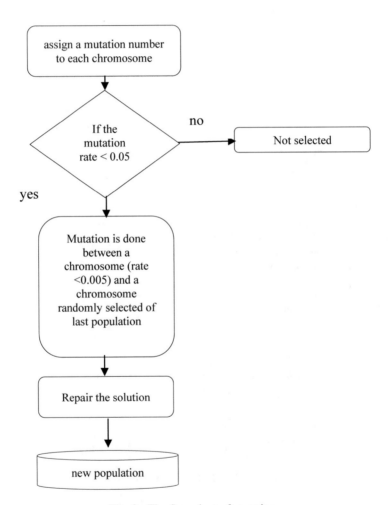

Fig. 3. The flow chart of mutation

5 Implementation

In this paper, the goal is to minimize the (sp) routing problem, the optimal solution is the shortest one, the step for compilation can generalized as:

Step1 : begin with the constraint limits is set for the sp route
  Population size=20;
  Number of iterations = 1000;
  Crossover rate =0.08;
  Mutation rate = 0.05;
Step 2: repeat for i =0 to i=length of database do chromosome randomly,
Step 3: evaluated each solution by calculating the distance and fitness between source node and destination node
Step 4: repeat for i=o to i= size of population do initial population (P) ;
Step 5: for each individual i ϵ P do calculate distance fd(i) and fitness ff(i);
  Then sort the population
  After that chooses the best solution = maxff(im);
Step 6: apply crossover:
  Repeat for i=0 to i=number of iterations,
        do for i=0 to i<i=population size, ia and ib ϵ P,
        if parent ((ia).random) < crossover rate,
        do select parent (ib) randomly from P,
        ic= do crossover (ia, ib);
        Repair(ic);
        if parent ((ic).random) < mutation rate ,
            do select (id) randomly from P;
            ie= do mutation (ic, id);
            repair(id);
            sort the population P;
            calculate fitness of the best solution from current
            population P   maxff(in),
            compare(id, im);
                if maxff(in) > maxff(im);
                replace in by im;
        end;
     until converged;
     end,

6 Simulation Results

6.1 Simulation Results for a Network with 20 Nodes

In this section, the proposed algorithm GA is compared with Chang Wook Ahn [9], the choice of this comparison is based on the topology used in his article, that allows to make a real study on his proposed topology, which is rare to find it in another article (Fig. 4).

In this example the population size is taken to be the same as the number of nodes in the network. Firstly, we compare the objective-function value returned by my algorithms and others, the line of my algorithm indicates the optimal path and it converge quickly than Chang Wook Ahn's Algorithm, and the proposed algorithm can easily find the best cost in 4 generation. we will increase the size of the network to evaluate the quality of our program.

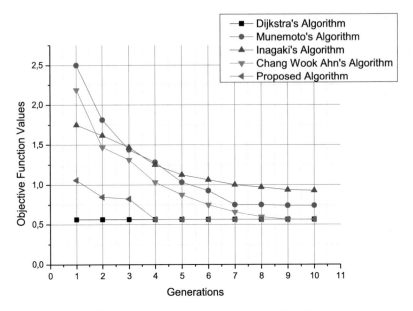

Fig. 4. Convergence property of each algorithm

6.2   Simulation Results for Random Network Topologies

In this paragraph, we generate some network topologies with 20 nodes to 1000 nodes randomly, for evaluate the quality of my program and the convergence speed, also we compare my algorithm proposed with Dijkstra's SP [12], know that, Dijkstra 's algorithm provides a reference point.

In this paragraph, we verify that the quality of solution, the convergence speed and the fitness function evaluations for each algorithm (Fig. 5).

Table 1. Performance comparison

Nodes	Algorithms		
	Dijkstra's Algorithm	Proposed Algorithm	
	Time(s)	Time(s)	Fitness
20	0,052	0,03	0.007
30	0,0954	0,051	0.0015
40	0,1698	0,0724	0.0007
50	0,3984	0,1054	0.0013
100	1,5915	0,4435	0.001
500	30,0132	1,1723	0.006
1000	127	4,2198	0.004

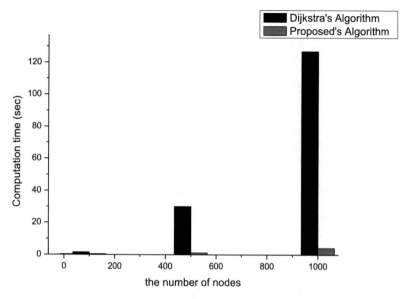

Fig. 5. Convergence performance between the Dijkstra's algorithm and proposed GA

From the above results we can say that the delay for my genetic algorithm is less than Dijkstra's algorithm mostly after 100 nodes. As illustrated in the Table 1 we calculate the fitness function for each network topologies from 20 to 1000 node, the fitness function must accurately measure the quality of the chromosomes in the population, so we notice that the solution quality as well as the execution time of the AG.

7   Conclusion

This paper proposes a genetic algorithm approach to the shortest path (SP) routing problem, every time when the GA start, a new individual will be produced, therefor, start the GA again and again the population will be constructed. AS well as, to guarantee the quality of an individual, pairs of chromosomes are selected to crossover and mutation, the experiment results show that it's feasible and successful to do in this way.

Nevertheless, there should be further research on how to implement the parallel programming in this case of network ad hoc.

References

1. Giri, A.K., Lobiyala, D.K., Katti, C.P.: Optimization of value of parameters in Ad-hoc on demand multipath distance vector routing using teaching-learning based optimization. In: 3rd International Conference on Recent Trends in Computing (ICRTC 2015), Elsevier (2015)

2. Holland, J.H.: Adaptation in Natural and Artificial Systems. The University of Michigan Press, USA (1975)
3. Goldberg, D.E.: Genetic Algorithms in Search, Optimization, and Machine Learning. Addison-Wesley, USA (1989)
4. Mardle, S., Pascoe, S.: An overview of genetic algorithms for the solution of optimization problems. Comput. High. Educ. Econ. 13(1), 16–20 (1999)
5. Michalewicz, Z.: Genetic Algorithms + Data Structures = Evolution Programs, 3rd edn. Springer, Heidelberg (1996). https://doi.org/10.1007/978-3-662-03315-9
6. Lakshmanaprabu, S.K., Shankar, K., Rani, S.S., Abdulhay, E., Arunkumar, N., Ramirez, G., Uthayakumar, J., et al.: An effect of big data technology with ant colony optimization based routing in vehicular ad hoc networks: towards smart cities. J. Cleaner Prod. 217, 584–593 (2019)
7. Bello-Salau, H., Aibinu, A.M., Wang, Z., Onumanyi, A.J., Onwuka, E.N., Dukiya, J.J.: An optimized routing algorithm for vehicle ad-hoc networks. Eng. Sci. Technol. Int. J. (2019)
8. Harrabia, S., Jaffar, I.B., Ghedira, K.: Novel optimized routing scheme for vanets. Procedia Comput. Sci. 98, 32–39 (2016)
9. Lerman, I., Ngouenet, F.: Algorithmes génétiques séquentiels et parallèles pour une représentation affine des proximités, Rapport de Recherche de l'INRIA Rennes - Projet REPCO 2570, INRIA (1995)
10. Ahn, C.W., Ramakrishna, R.S.: A genetic algorithm for shortest path routing problem and the sizing of populations. IEEE Trans. Evol. Comput. 6(6), 566–576 (2002)
11. Ali, K., Badreddine, S.: Algorithme génétique Université des sciences et de la technologie Houari Boumediene
12. Stalling, W.: High-Speed Networks: TCP/IP and ATM Design Principles. Prentice-Hall, Englewood Cliffs (1998)

Implementation and Evaluation of an Intrusion Detection System for IoT: Against Routing Attacks

Mohamed Khardioui[1(✉)], Abdelouahed Bamou[1],
My Driss El Ouadghiri[1], and Badraddine Aghoutane[2]

[1] IA Laboratory Science Faculty, Moulay Ismail University of Meknes,
Meknes, Morocco
khardioui@yahoo.fr, abdelbamou@gmail.com,
dmelouad@gmail.com
[2] TTI Team, Polydisciplinary Faculty of Errachidia,
Moulay Ismail University of Meknes, Meknes, Morocco
b.aghoutane@gmail.com

Abstract. The intensive growth of technology and devices connected to the Internet has made the Internet of Things (IoT) an essential element in all sectors of activity. It can be found in our watches, houses, cars, refrigerators, industrial machines, etc. Simply put, The Internet of Things is the future of technology that makes it easier to collect, analyze and distribute data that some person can implement them to achieve information or knowledge. Despite these advantages, this evolution suffers from a major security problem. This is due in particular to their heterogeneous nature, as well as the constraints of these objects (Memory, Processing Capabilities and limited energy…) are the main vulnerabilities of the IoT that are the origin of various attacks. Thus, many solutions have been developed to secure the IoT, but this remains insufficient because of the limitations of these mechanisms. This document is dedicated to the implementation and evaluation of an intrusion detection system (IDS) against attacks targeting the routing protocol in the IoT environment. The evaluation of the IDS is carried out with an emphasis on energy consumption, and detection rate. To achieve this, we have selected Cooja software as the simulator.

Keywords: Security · IoT · Intrusion detection system · IDS · RPL · Sinkhole

1 Introduction

Recently, the Internet of Things has appeared as a new paradigm that is revolutionizing the field of computer networks and telecommunications. The Internet of Things is an integral part of the Internet of the future [1], that be made up of a vast interconnection of all sorts of objects (other than computers and mobile phones) in our environment, such as vehicles, roads, cities, animals, etc., for an intelligent world and a considerable improvement in our way of life. Nevertheless, opening objects to the Internet presents a serious security problem, as these objects become remotely accessible by any malicious host on the Internet. The Internet of Things employs a range of technologies to create a

© Springer Nature Switzerland AG 2020
M. Ezziyyani (Ed.): AI2SD 2019, LNNS 92, pp. 155–166, 2020.
https://doi.org/10.1007/978-3-030-33103-0_16

communication channel between sensors, such as radio frequency identification (RFID), near field communication (NFC), etc. [2]. These sensors collect the information and send it to the base station via a wireless sensor network (WSN). Routing Protocol for Low-Power and Loss Networks (RPL) is one of the routing protocols mainly intended for the Internet of Things [1]. Data transmission and communication between the IoT devices should be secured and guaranteed end to end.

Ultimately, IoT is surrounded by different forms of routing attacks [3], such, as Sinkhole, Selective, Forwarding, Sybil and HELLO flood. These attacks cause disruption to the overall network and create more menaces. As a consequence, the security and protection of topology have become a difficult challenge for IoT. In this work, we propose a simulation of a trust-based intrusion detection system (IDS), to detect routing attacks in IoT networks. This IDS follows a centralized placement strategy. In particular, the border router operates the main detection module and other lightweight modules are deployed in the various nodes of the IoT network. Once the IDS mechanism identifies any intrusion, it will send an alert to the person in charge of security and the latter will take the necessary precautions without delay. The rest of this document is organized as follows: Sect. 2 describes the existing works on Routing attacks. Section 3 shows how the RPL protocol works and describes its different control messages. Then, Sect. 4 gives an overview of some attacks on the RPL routing protocol, the framework of our system will be presented in Sect. 5, Sect. 6 with 7 have been reserved for the implementation and evaluation of our IDS system. Finally, Sect. 8 concludes the paper.

2   Related Works

I found that the literature is very rich in works relating to the design and evaluation of an intrusion detection system (IDS) in the IoT environment. These works employ, a variety of techniques, according to Ghosal [4] who describe four different kinds of IDS, there are signature-based IDS, specification-based IDS, anomaly-based IDS, and hybrid-based IDS. Wallgren et al. [5] proposed a detection method for detecting different routing against RPL, including sinkhole attacks, wormhole attacks, and selective-forwarding attacks. This method is called RoVer, which represents role-based verification. The results showed that the proposed IDS performance and its effectiveness in terms of attack detection rate, the number of false positives and false negatives are better. Zarpelo et al. [6] present an overview IoT specific IDS, and detail aspects of the works that propose specific IDS schemes for the IOT or develop attack detection strategies for IoT threats, that could be incorporated into the IDS, and categorize the proposed IDS in the literature according to the following criteria: detection method, IDS placement strategy, security threat and validation strategy. Stephen et al. [7] propose an active detection method that is used to dynamically detect sinkhole attack for the Internet of Things, they announce, that the proposed method gives a low rate of false positives and false negatives. In the future, this method should be simulated with the cooja simulator and the results should be examined in different topological circumstances of the network.

3 Operation of Protocol RPL

Routing Protocol for Low-Power and Lossy Networks (RPL) is a proactive routing protocol based on a distance vector algorithm. It works like building an acyclic graph oriented and directed to a destination that is the edge router (DODAG: Destination Oriented Directed Acyclic Graph) [8]. Thus, it creates a routing tree with the edge router as its root. The roads in this tree are built according to a routing metric that allows nodes to assign themselves a rank corresponding to their location in the DAOD. As a result, the border router well must always take the lowest rank, as long as the other sensors increase their ranks as they descend into the DAOD. The construction of this tree is done through four ICMPv6 messages which are:

- DODAG Information Object (DIO) is a multicast message sent by each node and relayed by others to their neighbors, it allows a node to discover an RPL instance and join it through information from ascending routes.
- DODAG Information Solicitation (DIS) is used by nodes to request graph information from neighboring nodes that will respond by sending a DIO message.
- Destination Advertisement Object (DAO) is a unicast message sent by the nodes to propagate destination information to the top of the DAOD. This means that each node sends all its known descending routes to all its parents.
- Destination Advertisement Object Acknowledgement (DAO_ACK) is sent by the parent node to the child node in response to its DAO message (to acknowledge receipt).

The border router periodically sends a DIO message to the sensors within its radio range. This message contains information such as the rank of the sender and the objective function to be used [9]. This function is designed according to a routing metric. A sensor that receives a DIO message calculates its rank by adding its parent's rank to the cost of the metric. Then it sends this message to these neighbors. The DIO message can be requested by a DIS message. The sensor adds to its list of parents, the nodes with which it has had the smallest rows. The preferred parent is the one with who has had the lowest rank. It informs his parents of his choices by a DAO message. This message can be acknowledged by a DAO_ACK message.

4 Attacks Against RPL

The Protocol (RPL) is a standard routing protocol for constrained environments, such as IPv6 Low power Wireless Personal Area Networks (6LoWPAN). Guaranteeing the security of this type of network is a difficult challenge to meet, as devices are connected to the unreliable Internet and are limited in resources. However, this protocol is exposed to a wide variety of attacks [10]. Their consequences can be very heavy in terms of network efficiency and resources. In this section, we propose to establish a taxonomy of attacks against this protocol, taking into account three main categories [11], including attacks targeting the resource network, attacks modifying network topology and attacks related to network traffic.

4.1   Attacks Targeting Resources

Usually rest to have legitimate nodes perform unnecessary processing, exhaust their resources. This category of attacks targets energy consumption, and memory. It can have an impact on the availability of the network by cluttering up the available links and thus on the life of the network.

4.2   Attacks Against Traffic

This second category concerns attacks against RPL network traffic. It mainly includes the eavesdropping attack and the misappropriation attack.

4.3   Attacks on Topology

Attacks against the RPL protocol can also target network topology. We distinguish two main categories among these attacks: under-optimization and isolation.

- In case of sub-optimization attacks, the network will not converge to the optimal Form (i.e. optimal paths) resulting in poor performance.
- Attacks against topology can also isolate a node or a subset of nodes in the RPL Network, which means that these nodes are no longer able to communicate with their parents or with the root; this type presents the isolation attack.

In the following, we will summarize some of the attacks that target topology: see Fig. 1.

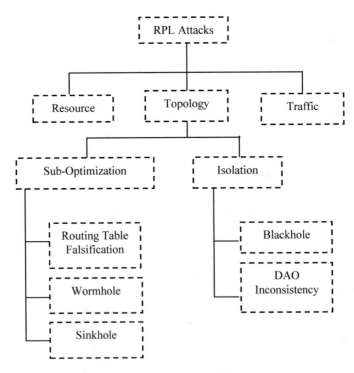

Fig. 1. Attacks against the RPL network

5 System Design

The Sinkhole attack creates more vulnerabilities than any other attack in (6LoWPAN) [12]. As, network protection is necessary, researchers are urged to develop an intrusion detection system (IDS), to mitigate the impact of these threats targeted the RPL routing protocol. This mechanism is designed to identify abnormal or suspicious activities on the target under analysis.

5.1 Illustration of the Problem

The sinkhole attack is an attack where the opponent attracts all the traffic to himself. In this attack, a malicious node extends its sphere of influence and tries to attract traffic from neighboring nodes. One way to perform a sinkhole attack in RPL would be for the malicious node to announce a lower rank than the node should have originally. This would cause each neighbor to choose the malicious node as a parent, and all traffic will be send through the node controlled by the attacker.

5.2 Intrusion Detection System

The intrusion detection solution for the sinkhole attack offers a detection model based on reputation and trust at each node [13], Each node calculates an index that models the reputation of the other nodes. The nodes then propagate their statuses, which are analyzed by the intrusion detection system. The reputation of each node is compared to a threshold, which determines whether or not the node is considered a node malicious.

The implementation of the method requires the use of DIO control messages, as there are different types of control messages in RPL for information exchange and topology maintenance. All these messages are sent as ICMPv6 requests.

One of these control messages is the DODAG Information Object (DIO), mainly used for routing control information. For example, it stores information on the IPv6 address of the roots. DIO messages carry information that allows a node to discover other nodes, a node rank, information about its configuration settings, and information about DODAG maintenance.

5.3 Description of the Proposed Algorithm

Our IDS is a trust-based intrusion detection system such that, each node in the network must observe and evaluate its neighbors based the routing information. These observations are sent to the border router, which analyzes them to announce whether the node is malicious or not [14]. The border router, in turn, receives trust messages from the nodes, processes the information they contain. Then, it updates the list that contains information about all DAODG nodes, as its trust observations. Based on this list, the router compares

Algorithm.1 . Pseudocode for Main Program – Border Router

```
etimer_set(&periodic, 0.1);
while(1) {
     PROCESS_YIELD();
     if(ev == tcpip_event) {
           print_data_received();
     }
     if(!(new_data < 0)) {
           isNodeMalicious(trustAddress,negative_value,
           positive_value);
           new_data = -1;
     }
     if(etimer_expired(&periodic)) {
           etimer_reset(&periodic);
     }
}
```

Algorithm.2 . Pseudocode for Main Program – Normal Node

```
route = uip_ds6_route_head();
while(route != NULL) {
   nexthop = uip_ds6_route_nexthop(route);
   if(uip_ipaddr_cmp(&from, nexthop)) {
           /* The sender of the DIO is a direct child
           * Test the rank of the node */
        if(currentNodesRank > DIOrank){
           /* The current node have a rank that is not
           * Allowed. Send negative observation to
           * the border router.*/
           tru_output(server_ipaddr, from, 0);
     } else {
           /*The current rank is correct
           * Send positive observation to border router*/
           tru_output(server_ipaddr, from, 1);
     }
   }
   route = uip_ds6_route_next(route);
}
```

[Pseudo code from: https://github.com/freddyny/Master---IDS].

6 Experimentation

This section presents the results of the evaluation of our IDS against the sinkhole attack. Next, we briefly present the state of the art of the most relevant IDS evaluations and analyze them to find out what is wrong with these evaluations. At this stage, we can identify their weaknesses and therefore specify the requirements for a satisfactory evaluation of the IDS. A detailed description of the network configuration will also be presented for each experiment and the corresponding parameters that are used in the Cooja simulator.

6.1 Energy Consumption Profile

- Experiment 1: Energy consumption without IDS
  The purpose of this experiment is to deduce how much of the energy consumed by the nodes and border router before implementing the IDS. The experiment consists of seven nodes, where one node is a border router and the rest of the nodes are non-malicious nodes. The network contains no attackers because the goal is to see how many resources the RPL network uses without the IDS.
- Experiment 2: Energy Consumption with IDS
  The determination of the energy consumed by the network's components after the IDS implementation is one of the objectives of this experiment. As well, the comparison between the results obtained in experiments one and two allows us to announce that the deviation of the energy consumed before and after the implementation of the IDS is acceptable or not (Fig. 2) (Table 1).
- Results
  From the graph above, it is clear that the energy consumed by the normal nodes after the implementation of the IDS is almost the same as the energy consumed before its implementation (Fig. 3).
  For the router the same comment with an increase in energy consumption after the IDS implementing. The following table summarizes the results obtained (Table 2).

Table 1. The following table shows the configuration of the previous experiments

	Experiment 1	Experiment 2
Number of attackers	0	1
Number of normal nodes	6	6
Border router	1	1
Experiment time	4 h, 50 min	4 h, 50 min

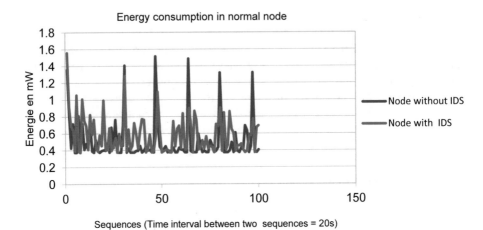

Fig. 2. A graph that shows energy consumption in the normal node before and after the IDS implementing

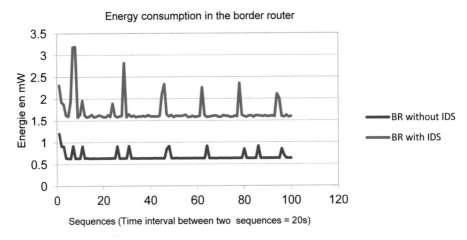

Fig. 3. A graph that shows energy consumption in the Border Router before and after the IDS implementing

Table 2. Shows energy consumption in the different devices

Devices	Energy consumption without IDS	Energy consumption with IDS
Border router	286.40 mW	572 mW
Normal node	232 mW	243 mW

6.2 IDS Evaluation: Real Alarms and Time to Detection

This paragraph's goal is the calculation of the detection rate of our IDS, and the time required to detect all attacks. So, we have decided to carry out the following experiments

- Experiment 3:
  The configuration of this experiment is as follows, ten nodes, and border router that is always in listening mode, and four attackers. Each attacker carries out a sinkhole attack. The time of the experiment is 30 min. The first attack is introduced after three minutes of the experiment beginning, and each attack among the remains will be introduced after the same waiting time.
- Experiment 4:
  The configuration of this experiment is as follows, ten nodes, and border router that is always in listening mode, and six attackers. Each attacker carries out a sinkhole attack. The time of the experiment is 1 h, and the intrusion of two successive attacks, separated by three minutes (Table 3).

Table 3. A table showing the number of attackers detected, false positives, false negatives and the time before detecting the first and last attack

	Experiment 3	Experiment 4
Attackers detected	4	5
False positives	2	4
False negatives	0	1
Time before detecting the first attack	5 min	9 min
Time before detecting the last attack	17 min	1 h

- Results
  In this table, False Positive is an event that occurs when a system classifies an action as abnormal when it is a legitimate action; False Negative is an event that occurs when an intrusion is identified as non-intrusive by the system and lets it happen.
  It can be seen that during the two experiments, 6 false alarms were triggered, and almost 91% of the attacks were detected
  The obtained data is used to calculate the criteria that determine the system's quality. And here are some of them:

$$\text{The false negative rate} = \frac{\text{false negative}}{\text{false negative} + \text{true positive}} = 16\% \quad (1)$$

$$\text{The false positive rate} = \frac{\text{false positive}}{\text{false positive} + \text{true negative}} = 38\% \quad (2)$$

$$\text{The detection rate} = \frac{\text{true alarms}}{\text{launched attacks}} = 91\% \quad (3)$$

7   Discussion

7.1   Energy Consumption Profile

It is necessary to take into account energy consumption. As the devices that will use the IDS could be deployed in a difficult to access space, and they are also devices powered by a limited capacity power source (battery).

From the previous results, we will deduce that our system is very efficient that it has a very limited negative impact on the performance of the monitored objects. Indeed, the energy consumption by a normal node after the implementation of IDS is almost the same energy consumed without IDS. On the other hand, our system dragged a significant increase in the energy consumed for the border router. However, experiments have shown that energy consumption is increased by 329 mW in 4 h, 50 min compared to the normal state (without IDS). This may be justified as most of the processing requested by the IDS, will be carried out in the embedded router, which is in charge of receiving, analyzing and storing observations. However, the only task of a normal node is sending the observations about its neighbors toward the border router.

7.2   False Alarms

As indicated previously, IDS induces a high number of false positives. This is due owing to the way the intrusion detection system detects the sinkhole attack. The high number of false positives in Experiment three and four can be explained by the fact that: The IDS can give negative observations to non-malicious nodes. However, the number of positive observations must be sufficient so that these nodes are not marked as malicious. The router may receive more negative than positive comments, as the positive observations are lost due to network flooding, causing the router to declare a non-malicious node to be malicious.

Three other reasons for the IDS to introduce false positives:

- Small differences in the rank of two non-malicious nodes.
- Non-malicious nodes that are calculating their rank based on the rank of the malicious nodes. Thus an incorrect rank then it receives negative observations.
- The implementation could be wrong.

7.3   Detection Rate

Experiments have shown that the detection rate of our system is 91%. This is a significant detection rate, but it still results in enough false positives. This rate can reach 100%, but with a prolonged detection time. However, this will probably depend on the configuration of the network used, one of the most important factors being the number of nodes that can join the attack node and choose it as a parent.

The time, it takes for an attack to be detected is important for an IDS, because, if the attack spends a lot of time in the system. It can be harmful to it because of the amount of damage it can cause. Experiments 3 and 4 show how long it has taken for the attacks to be detected. While the experiment shows that the IDS spends an average of 12 min to detect attacks, the time is longer when you have more nodes or more attackers to detect an attack. This is because the edge router does not receive all negative observations, and it takes longer before a negative observation is received. The router must be in listening mode at all times. This requires more energy consuming on the part of the router, but also significantly improves the results.

8 Conclusion

In this document, we propose the steps to follow in order to implement and evaluate an intrusion detection system (IDS) in the IoT environment. This system has a goal, the detection of the Sinkhole attack against the RPL routing protocol. The detection technique used by our system is based on the signature, and the trust placed in each node. The chosen platform for developing and testing the IDS solution is the Cooja simulator, which supports application development for Contiki OS.

According to the results reported in previous experiments, it was noticed and concluded that our solution can detect intrusions on connected objects efficiently and with a high detection rate. So, the system has a very limited impact on the performance of the monitored objects. As, we have already pointed out that our system, it has some weaknesses. Those are shown in an increase in the energy consumed by the border router, as well as a high rate of false positives.

In our future work, we will also evaluate the performance of the IDS by reducing false positives during the attack detection process and extending the IDS to detect other routine attacks. Finally, we will import the IDS modules into Contiki OS to test performance in a real IoT environment.

References

1. Atzori, L., Iera, A., Morabito, G.: Understanding the Internet of Things: definition, potentials, and societal role of a fast evolving paradigm. Ad Hoc Netw. **56**, 122–140 (2017). https://doi.org/10.1016/j.adhoc.2016.12.004
2. Li, S , Tryfonas, T., Li, H.: The Internet of Things: a security point of view. Internet Res. **26**, 337–359 (2016). https://doi.org/10.1108/IntR-07-2014-0173
3. Khan, M.M., Lodhi, M.A., Rehman, A., Khan, A., Hussain, F.B.: Sink-to-sink coordination framework using RPL: routing protocol for low power and lossy networks. J. Sens. **2016**, 1–11 (2016). https://doi.org/10.1155/2016/2635429
4. Ghosal, A., Halder, S.: A survey on energy efficient intrusion detection in wireless sensor networks. J. Ambient Intell. Smart Environ. **9**, 239–261 (2017). https://doi.org/10.3233/AIS-170426
5. Wallgren, L., Raza, S., Voigt, T.: Routing attacks and countermeasures in the RPL-based Internet of Things. Int. J. Distrib. Sens. Netw. **9**, 794326 (2013). https://doi.org/10.1155/2013/794326
6. Stephen, R.: Deist: dynamic detection of Sinkhole attack for Internet of Things. Int. J. Eng. Comput. Sci. **5**(12), 19358–19362 (2016). https://doi.org/10.18535/ijecs/v5i12.16. Research scholar Department of Computer Science St. Joseph's College (Autonomous) Tiruchirappalli-620002

7. Colom, J.F., Gil, D., Mora, H., Volckaert, B., Jimeno, A.M.: Scheduling framework for distributed intrusion detection systems over heterogeneous network architectures. J. Netw. Comput. Appl. **108**, 76–86 (2018). https://doi.org/10.1016/j.jnca.2018.02.004

8. Clausen, T., Herberg, U., Philipp, M.: A critical evaluation of the IPv6 routing protocol for low power and lossy networks (RPL). In: 2011 IEEE 7th International Conference on Wireless and Mobile Computing, Networking and Communications (WiMob), pp. 365–372. IEEE, Shanghai (2011)

9. Vucinic, M., Romaniello, G., Guelorget, L., Tourancheau, B., Rousseau, F., Alphand, O., Duda, A., Damon, L.: Topology construction in RPL networks over Beacon-enabled 802.15.4. ArXiv arXiv:1404.7803 Cs (2014)

10. Pu, C., Song, T.: Hatchetman attack: a denial of service attack against routing in low power and lossy networks. In: 2018 5th IEEE International Conference on Cyber Security and Cloud Computing (CSCloud)/2018 4th IEEE International Conference on Edge Computing and Scalable Cloud (EdgeCom), pp. 12–17. IEEE, Shanghai (2018)

11. Mayzaud, A., Badonnel, R., Chrisment, I.: A taxonomy of attacks in RPL-based Internet of Things. vol. 16 (2016)

12. Sundararajan, R.K., Arumugam, U.: Intrusion detection algorithm for mitigating sinkhole attack on LEACH protocol in wireless sensor networks. J. Sens. **2015**, 1–12 (2015). https://doi.org/10.1155/2015/203814

13. Ioulianou, P., Vasilakis, V., Moscholios, I., Logothetis, M.: A signature-based intrusion detection system for the Internet of Things. 7 (2018)

14. Nygaard, F.: Intrusion detection system in IoT. 125 (2017)

Implementing and Evaluating an Intrusion Detection System for Denial of Service Attacks in IoT Environments

Abdelouahed Bamou[1]([⊠]), Mohamed Khardioui[1],
My Driss El Ouadghiri[1], and Badraddine Aghoutane[2]

[1] IA Laboratory Science Faculty, My Ismail University of Meknes,
Meknes, Morocco
abdelbamou@gmail.com, khardioui@yahoo.fr,
dmelouad@gmail.com,
[2] Team of Processing and Transformation of Information, Polydisciplinary
Faculty of Errachidia, Moulay Ismail University of Meknes, Meknes, Morocco
b.aghoutane@gmail.com

Abstract. The Internet of Things (IoT) is very successful in different fields: industry, health, logistics, smart cities, smart home. Despite this success, this new technology suffers from a big security problem. Much of this problem is due to the constraints of connected objects (Memory, Processing Capabilities and limited energy ...) that are the cause of various attacks, mainly denial of service attacks (DOS/DDOS). In this paper we propose the implementation and evaluation of a system of intrusion detection DOS attacks, based on the verification of the abnormal use of the energy consumption of connected objects in IoT environments. The implementation of the proposed algorithm is carried out in the Contiki-Coja environment and simulation results indicate that denial-of-service attacks can be detected with high accuracy, while keeping the number of false-positives very low.

Keywords: DOS · Security IoT · Intrusion Detection System (IDS)

1 Introduction

The Internet of Things (IoT) has flourished in many critical applications. Smart devices are gaining popularity in different sectors, such as environmental monitoring, medical treatment and public health, smart home, smart grid, smart cities and in many other areas [1] with the promise to make our lives easier and more comfortable. According to a prediction by Cisco's Internet of Things Group, there will be over 50 billion connected devices by 2020 [2]. IoT enables objects and processes in real-time to be monitored, linked and interacted [1]. However, the increased deployment of such smart devices brings an increase in potential security risks. IoT devices do not only collect users' personal information, but also could acquire users' activity, which could lead to huge security issues [3]. The devices and services suffer from intrusions, attacks and malicious activities. To protect the IoT network against known and unknown attacks,

© Springer Nature Switzerland AG 2020
M. Ezziyyani (Ed.): AI2SD 2019, LNNS 92, pp. 167–178, 2020.
https://doi.org/10.1007/978-3-030-33103-0_17

the better solution is Intrusion Detection System (IDS) to match with the characteristics of IoT, which requires real-time monitoring.

An IDS is a security technology pristinely built for detecting various exploits against a target application or computer. He could detect, usually via alerts, malicious activities and suspicious transactions [4, 5].

The performance of an intrusion detection system, and in particular its analysis method is related to two important notions that make it possible to evaluate these performances [6]:

The reliability (or coverage) of an Analyzer is related to its false negative rate, which represents the percentage of undetected intrusions, this rate should be as low as possible.

The relevance (or credibility) of an Analyzer is related to its false positive rate, which represents the percentage of false alerts.

There are three methods to categorize IDS [7, 8]:

- Host-based vs Network –based:
  This division of the IDS is based on the place where it is mounted. When an IDS is placed on the network segment it is said to be a Network Intrusion Detection System, whereas, when an IDS is deployed in workstations, they are said to be Host-based Intrusion Detection Systems.
- Active system vs Passive system:
  The active IDS respond to the suspicious activity; the passive IDS detect attack, log the information and signal an alert.
- Misuse detection vs Anomaly Detection:
  The misuse detection, also known as signature-based method, seems for a specific signature to match, signaling an intrusion. They can detect many or all known attack patterns, but the weakness of signature based intrusion detection systems is the incapability of identifying new types of attacks or variations of known attacks.

The anomaly detection was originally proposed in [9]. In anomaly based IDSs, the normal behavior of the system or network traffic are represented and, for any behavior that varies over a pre-defined threshold, an anomalous activity is identified. Detecting the attacks according to the anomalies that occur in the network amounts to establishing, as a first step, a model of the normal operation of the network. In a second step, the monitoring itself is triggered, and if the observed variables substantially derogate from the generated model; an alert is raised [10]. These methods therefore make it possible to intervene when a problem actually occurs in the network, without its origin having necessarily been identified. They have the advantage of adapting to new types of attacks, for which they had not been imagined at the beginning. However, an important issue in anomaly based IDSs is how these systems should be trained, i.e., how to define what is a normal behavior of a system or network environment (which features are relevant) and how to represent this behavior computationally; the second issue is the number of false positives generated are higher than on those based on signatures.

Anomaly-based detection groups three subcategories [11]:

- The first is that of statistical models, which initially record series of values for variables related to the proper functioning of the network, and then seek to identify significant deviations from these reference values. The observed variables, directly related to traffic, can be considered as independent variables according to Gaussian laws, as correlated variables, or as time series.
- A second group models based on modeled knowledge of normal network operations, and uses a model generated according to data and using tools such as description languages or finite automata.
- Finally, detection can be based on learning methods, where models are created and updated via formal tools such as Markov processes, Bayesian networks, fuzzy logic, genetic algorithms, neural networks,

Latest security incident trend statistics, currently showing an increase in denial of service (DoS) attacks, stress out availability as the security objective of choice when it comes to malicious online activities. A Dos attack is a type of attack designed to make an organization's services or resources unavailable for an indefinite time [12]. He has been the biggest threat to the Internet and can cause heavy losses for businesses and governments. As these attacks are typically performed on a massive scale using a large number of computer resources, they are commonly referred to as distributed denial of service attacks (DDoS). With the development of emerging technologies, such as the Internet of Things, attackers can launch DoS attacks at a lower cost and it becomes much more difficult to detect and to prevent [1].

DoS attacks, in general, execute some form of direct or indirect resource exhaustion on target side through specific traffic generation, thus performing disruption with a negative impact on service readiness and continuity. Such events present a serious threat to security, dependability and their intersections [12].

The purpose of this paper is the implementation and evaluation of an intrusion detection mechanism of DOS attacks adapted to connected objects by using the learning method. We propose an efficient and low cost resource solution based on energy consumption as a metric to detect attacks

The implementation of the IDS is done in the Contiki operating system [13] and tested in the Cooja simulation environment.

We present in Sect. 2 the related work, in Sect. 3 we expose the algorithm of the DOS attack detection based on the energy consumption of the nodes, in Sect. 4, we present the evaluation metrics then the procedure and the discussion of the results, we conclude on the main results and perspectives.

2   Related Work

Multiple IDSs for Iot have been proposed. Existing systems differ not only in architecture, but also in applied detection techniques and detectable attacks. Some are not limited to particular malicious behavior, while others can only detect specific attacks such as sinkholes or wormholes.

Cho et al. [11–14], proposed a detection scheme for botnets using the anomaly-based method. The authors supposed that botnets cause unpredicted changes in the traffic of 6LoWPAN sensor nodes. The proposed solution calculates the average for three metrics to compose the normal behavior profile: packet length, the sum of TCP control field, and the number of connections of each node. Then, the system monitors network traffic and display an alert when metrics for any node violate the computed averages.

Raza et al. [15] proposed an IDS, called SVELTE, to detect sinkhole and selective forwarding attacks. The objective of this hybrid IDS is to offer a satisfactory tradeoff between storage cost of the signature-based method and computing cost of the anomaly-based method.

Pongle and Chavan [16] presented an IDS designed to detect wormhole attacks in IoT devices. The authors assume that the wormhole attack always leaves its symptoms on the system, for example, a high number of control packets are exchanged between the two ends of the tunnel, or a high number of neighbors get formed after a successful attack. Using this logic, the authors propose three algorithms to detect such anomalies in the network. According to their experimentation, the system achieved a true positive rate of 94% for wormhole detection and 87% for detecting both the attacker and the attack. The authors also performed a study on power and memory consumption of the nodes.

Some authors have already used energy consumption as a metric for intrusion detection. Nash et al. [17] presented an IDS for mobile computers that uses several parameters like CPU load and disk read and write access to estimate the power consumption.

Shen et al. [18] proposed an intrusion detection scheme comparing the energy consumptions of different sensor nodes in clustered wireless sensor networks (WSNs). It can differentiate the types of DoS attacks such as selective forwarding and wormhole attacks with the assistance of energy thresholds. Nodes are required to send messages containing their remaining energy to the base station. The energy consumption of all nodes is then predicted at the base station using Markov chains. The scheme looks for nodes, which spend significantly more energy than the other nodes.

Lee et al. [19] suggested an IDS based on analyzing energy consumption of nodes to detect possible DoS attacks. They defined models of regular energy consumption for mesh-under routing scheme and route-over routing scheme. Then, each node monitors its energy consumption at a sampling rate of 0.5 s. When the energy consumption deviates from the expected value, the IDS classify the node as malicious and remove it from the route table in 6LoWPAN. The authors claimed that it is a lightweight approach, specifically developed for low capacity networks.

Riecker et al. [20] propose an architecture of a lightweight intrusion detection system that is based on energy readings and linear regression. They use a moving agent which moves around from node to node in the network like a token to collect readings of energy consumption by the nodes. The mobile agent uses the energy readings collected to make predictions on how much energy the node should have been consuming the next time the agent arrives based on linear regression models. Their work also serves as a demonstration that power consumption is a suitable metric for detecting denial-of-service-attacks with high accuracy while maintaining low false-positive rates.

3 DOS Attack Detection Based on Node Energy Consumption

The solution proposes to measure the energy consumption to see if a node is attacked or not. It collects the history of the node's energy consumption values and compares the change in average energy consumption. The method calculates an average over a specified history size that can be changed as a parameter to find the optimal value. If the calculated average values increase for a specified number of consecutive turns, the node reports that it is being attacked. When an attack occurs, the last values will be replaced by the oldest, lower ones. This allows you to generate continuous alarms as long as the attack is in progress.

Indeed, when the attack is detected, we replace the current list by the list of normal values already recorded, in this way the comparison of the current energy values is done with the normal state during the next turn, and not with the values taken during the attack.

The same procedure is repeated to repair the second anomaly, but this time by injecting the normal values each time the current tower energy average is lower than the previous lap.

The detection algorithm of the method based on energy consumption:

```
Algorithm Energy_Increasing

Require: Time_interval : T

Require: increasing_Energy_limit : L

Require:  Normal_Energy_Consumption  in  interval  T  (without  attacks):
Energy_Trust

increasing_Energy_counter ← 0

Repeat for ever

Energy_value ←  Energy_Consumption in interval T

   if    Energy_value > Energy_Trust then

    increasing_Energy_counter ++

         if      increasing_Energy_counter > L    then

               Send Alarm()

         end if

   else

 increasing_ Energy _counter ← 0

   end if

end repeat.
```

4   Rating Metrics

The detection methods of our IDS are evaluated according to the following metrics:

4.1   Energy Consumption

The total energy consumption of the node is measured by calculating the energy consumption of the communication and computations, as described in the equation [10]:

$$Etot = Eradio + ECPU$$

The calculation of the time devoted to the communication is done by measuring the transmission and listening times of the radio module. These times are then used to calculate energy consumption for radio activities according to the equation:

$$Eradio = (TReceive \cdot IRadio\,RX + TTransmit \cdot IRadio\,TX) \cdot VSystem$$

Where TReceive is the time spent receiving radio communications, TTransmit is the transmission time, VSystem is the operating voltage of the system, IRadio RX is the power consumption required for reception, and IRadio TX is the power consumption required for the transmission. IRadio RX, IRadio TX and VSystem are extracted from the sensor node manuals used.

The CPU power consumption can be calculated by measuring the time spent on the calculations and the time spent in sleep mode, and then using these values to calculate the total power consumption of the CPU according to the equation:

$$ECPU = (Active\,TCPU \cdot Active\,ICPU + TCPU\,Standby \cdot ICPU\,Standby) \cdot VSystem$$

4.2   Detection Rate

The detection rate is an important measure. It is calculated by taking the number of true alarms generated by the IDS and dividing it by the number of attacks launched against the network, as shown by the equation:

Detection rate = the number of true alarms/the number of attacks launched.

The number of false alarms is the number of erroneous alarms generated by the IDS.

4.3   The Time Elapsed Until the Detection (TUD)

It measures the time needed for the IDS to detect an attack from the moment the attack is launched:

$$TUD = Tdetected - Tlaunched$$

5 Experimentation

We run our experiments in Contiki's network simulator Cooja that has shown to produce realistic results [13]. Cooja runs deployable Contiki code. In our simulations, we use emulated Tmote Sky nodes.

The first test measures the power consumption while the node is performing both the detection method and a simple program that periodically sends sensor values to the receiving node (Fig. 2). It serves to show what a normal power consumption of the sensor node looks like and serves as a basis for comparison with the variation in power consumption when the node is attacked.

The other tests measure power consumption when the node executes both a program sending sensor values to the receiving node and the detection method (Fig. 2) while undergoing attacks with varying degrees of intensity (Fig. 1).

Fig. 1. Configuring the IDS experience based on energy consumption

Table 1. Shows the intensities of DOS attacks

Intensity name	Number of messages sent per second
Low	1.28
Medium low	2.56
Medium high	5.12
High	128

The table describes how often the attacking node attempts to send messages in terms of messages per second. In the case of high intensity for example, the node will try to send 128 messages every second as shown in the Table 1.

The attack is ongoing throughout the test interval and the results serve to demonstrate the viability of the detection method.

6 Results

6.1 Energy Consumption

Figure 2 displays the resulting graphs of the energy consumption of the node when he is running both the example program and the detection method while being under attacks with varying degrees of intensities:

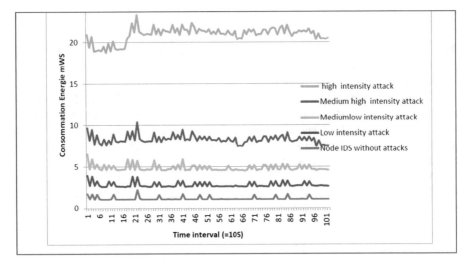

Fig. 2. Comparing the energy consumption of a node according to different intensities of attacks

Looking at Fig. 2, it can be noted that the energy consumption increases according to the intensity of the attacks as expected. For low intensity, all periods have an energy value of more than 1.50 mws, eight periods of more than 2.10 mws. For the medium low intensity, it can be noted that the energy increases so that all the values are greater than 3.94 mws and that there are peaks in the energy consumption. For the medium high intensity attack, it can be noted that all periods have an energy value of at least 7 mws with some peaks reaching more than 8,50 mws. For the high intensity attack, it can be noted that the energy values for all periods reach more than 18 mws and many reach about 22 mws.

By comparing the graph elements (Fig. 2), we can conclude that the energy consumption is not regular and linear for the chosen duration. The pattern of power consumption when read once a tenth of a second has many peaks and variations. However with the use of our algorithm, one or two peaks do not cause the node to alarm since the average must increase for a number of consecutive turns before any alarm is triggered.

6.2   Parameters for the Detection Algorithm

In order to adjust the performance of the DoS attack detection method and to avoid giving false alarms, the parameters for the frequency of reading of the energy (Time_interval) and increasing energy limit must be decided. To do this, different tests are designed with varying values for the above mentioned parameters and the results give the following recommendations:

Time_interval, T = 10 s, increasing energy limit, L = 5.

The data presented for the chosen configuration, allow to determine the parameter for how many rounds the value can continue to increase before sending an alarm (5 consecutive revolutions), for an energy reading frequency of 10 s.

6.3 Results in Terms of False Alarms, Detection Rate and Time Until the Detection of the Attack

Table 2. Shows the (averaged over 10 test-runs) time until detection for all the attacks that are discovered or not detected which are marked with N/D. during all the 10 tests (where the each test is run for 5 min).

True alarms and Time Until Detection (TUD) (seconds)				
Delay	Intensity			
	Low	Medium low	Medium high	High
Short	N/D	N/D	120	60.01
Medium short	220	60	54	54.01
Medium long	180	57	50.04	50.08
Long	123	55	50.06	40.53

It can be seen that only one of the four low intensity attacks is not detected but with a time until the detection long. The low-medium attack with the short delay is not detected and all other attacks are detected.

The data obtained is used to calculate the false alarm rate:

False positive rate = 0

False negative rate = 2/16 = 0.125 = 12.5%

The Table 2 is used to calculate the detection rate.

The Equation:

$$\text{Detection Rate Attacks} = 14/16 = 0.875 = 87.5\%$$

Shows the detection rate with low intensity attacks included, but it should be noted that low intensity attacks need a long time to high detection.

7 Discussion

7.1 Energy Consumption

Figure 2 showed that the energy consumption increases according to the intensity of the attacks, so it is a suitable metric to detect Dos attacks.

The difference in energy consumption between just running the example program and also running the IDS is as low 1 mJ for the algorithm while the system is not under an attack. The overhead produced by the IDS is relatively low. The small addition to the energy consumption can be seen as quite low in contrast to gaining protection against attacks Dos.

The limitation of the algorithm is that it does not detect low-intensity DoS attacks as well as those that start with a short delay, before the IDS fills its list with normal values.

7.2   Parameters for the Detection Algorithm

The behavior in a simulator and in the real world may differ a bit, so it is possible that other parameters were chosen if the initial test had also been done on real hardware.

In real scenarios, the energy consumption is not flat. In fact, the energy consumption varies a lot depending on what the node is currently doing. This problem can be overcome by choosing good parameters.

7.3   Detection Rate

For our IDS only two attacks are not detected but with difficulties to detect low intensity attacks. During these low intensity attacks, no disturbance of normal network function was observed. The nodes were able to send their messages with no apparent delay, and the receiving node received them as usual. It is important to determine when something counts as an attack when evaluating the method's detection rate for the algorithm. If the threshold for something to count as an attack is set to the medium low intensity (when the attacking node tries to send about two messages per second) the detection rate will increase from 87.5% to 92%. This is a significant increase in the result.

By comparing the node energy containing the IDS with and without the low intensity attack, it can be argued that the low intensity attack should be excluded to be considered an attack. This conclusion stems from the fact that the energy values for the two cases differ little. The detection algorithm would need to adjust its lowest settings to detect this type of attack, which would have high false positive rates.

The number of false alarms is low for the algorithm. There is no false positive in the test. The intensity of the attack in this case does not matter since the false alarms occur before launching the attack. However, it is not surprising that false alarms occur during the short delay and low intensities since the IDS has trouble detecting this type of attack as explained above.

8   Conclusion

We presented through this paper the implementation of IDS that uses the anomaly-based technique to detect DOS attacks, as well as the performance evaluation results of the proposed solution to ensure the security of the IoT.

The results obtained clearly show that energy consumption can be used as a basis for detecting DOS attacks. The algorithm is not only functional but also effective for detecting attacks with a high detection rate, while maintaining a zero false positive number throughout the evaluation. However the method is in difficulty for attacks of low intensities or short attack time.

The assessments are made using the Cooja network simulator of the Contiki operating system which is dedicated to the simulation of networks in the context of the IoT.

This IDS has shown promising results. However, there is still work to be done to get it ready to be implemented in a real IoT environment.

Indeed the IDS has limitations, the weakness of the IDS based on power consumption is that it does not detect the low intensity DoS attacks as well as the one that start with a short delay, before the IDS does not fill its list with normal values.

As prospects, we plan to integrate the proposed solutions into real applications of IoT.

References

1. Zhao, K., Ge, L.: A survey on the internet of things security. In: 2013 Ninth International Conference on Computational Intelligence and Security, Leshan, China (2013)
2. Cisco White Paper: The Internet of Things How the Next Evolution of the Internet Is Changing Everything, April 2011
3. Yang, Y., Wu, L., Yin, G., Li, L., Zhao, H.: A survey on security and privacy issues in internet-of-things. IEEE Internet Things J. $PP(99)$, 1 (2017)
4. Wagh, S.K., Pachghare, V.K., Kolhe, S.R.: Survey on intrusion detection system using machine learning techniques. Int. J. Comput. Appl. $78(16)$, 30–37 (2013)
5. Scarfone, K., Mell, P.: Guide to Intrusion Detection and Prevention Systems (IDPS). National Institute of Standards and Technology, Gaithersburg, MD 20899-8930, February 2007
6. Briffaut, J.: Formalisation et garantie de propriétés de sécurité système: application à la détection d'intrusions. Autre [cs.OH]. Université d'Orléans (2007)
7. Chebrolu, S., Abraham, A., Thomas, J.P.: Feature deduction and ensemble design of intrusion detection systems. Comput. Secur. $24(4)$, 295–307 (2005)
8. Wagh, S.K., Pachghare, V.K., Kolhe, S.R.: Survey on intrusion detection system using machine learning techniques. Int. J. Comput. Appl. $78(16)$ (2013). ISSN: 0975 – 8887
9. Denning, D.: An intrusion-detection model. IEEE Trans. Softw. Eng. 2, 222–232 (1987)
10. Becker, J., Vester, M.: Intrusion Detection System Framework for Internet of Things. Construction of an Intrusion Detection System for Wireless Sensor Nodes with Modern Hardware. Master's thesis in Computer Systems and Networks. Supervisor Olaf Landsiedel. Chalmers University of Technology, Gothenburg, Sweden, 120 pages (2017)
11. Zarpelãoa, B.B., Mianib, R.S., Kawakania, C.T., de Alvarengaa, S.C.: A survey of intrusion detection in internet of things. J. Netw. Comput. Appl. 84, 25–37 (2017)
12. Zlomislic, V., Fertalj, K., Sruk, V.: Denial of service attacks, defences and research challenges. Cluster Comput. 20, 661–671 (2017). https://doi.org/10.1007/s10586-017-0730-x
13. Thingsquare. Contiki: The Open Source OS for the Internet of Things (2017). http://www.contiki-os.org/. Accessed 01 Apr 2018
14. Cho, E., Kim, J., Hong, C.: Attack model and detection scheme for botnet on 6LoWPAN. In: Hong, C., Tonouchi, T., Ma, Y., Chao, C.-S. (eds.) Management Enabling the Future Internet for Changing Business and New Computing Services. Lecture Notes in Computer Science, vol. 5787, pp. 515–518. Springer, Heidelberg (2009)
15. Raza, S., Wallgren, L., Voigt, T.: SVELTE: real-time intrusion detection in the internet of things. Ad Hoc Netw. $11(8)$, 2661–2674 (2013)

16. Pongle, P., Chavan, G.: Real time intrusion and wormhole attack detection in internet of things. Int. J. Comput. Appl. **121**(9), 1–9 (2015)
17. Nash, D.C., Martin, T.L., Ha, D.S., Hsiao, M.S.: Towards an intrusion detection system for battery exhaustion attacks on mobile computing devices. In: Proceedings of the Third IEEE International Conference on Pervasive Computing and Communications Workshops, pp. 141–145 (2005)
18. Shen, W., Han, G., Shu, L., Rodrigues, J., Chilamkurti, N.: A new energy prediction approach for intrusion detection in cluster-based wireless sensor networks. In: Green Communications and Networking. volume 51 of Lecture Notes of the Institute for Computer Sciences, Social Informatics and Telecommunications Engineering, pp. 1–12. Springer, Heidelberg (2012)
19. Lee, T.-H., Wen, C.-H., Chang, L.-H., Chiang, H.-S., Hsieh, M.-C.: A lightweight intrusion detection scheme based on energy consumption analysis in 6LowPAN. In: Huang, Y.-M., Chao, H.-C., Deng, D.-J., Park, J.J.J.H. (eds.) Advanced Technologies, Embedded and Multimedia for Human-centric Computing. Lecture Notes in Electrical Engineering, vol. 260, pp. 1205–1213. Springer, Netherlands (2014)
20. Riecker, M., Biedermann, S., El Bansarkhani, R., Hollick, M.: Lightweight energy consumption-based intrusion detection system for wireless sensor networks. Int. J. Inf. Secur. **14**(2), 155–167 (2015)

Intelligent Urban Transport Decision Analysis System Based on Mining in Big Data Analytics and Data Warehouse

Khaoula Addakiri[1], Hajar Khallouki[2(✉)], and Mohamed Bahaj[2]

[1] Computer Science Department, Ibn Zohr University, Ouarzazate, Morocco
[2] Mathematics and Computer Science Department, Faculty of Sciences
and Technologies, Hassan I University, Settat, Morocco
Hajar.khallouki@gmail.com

Abstract. This paper conduct a study on the augmentation of the current capabilities of the intelligent urban mobility and road transport in terms of the analytics dimension focusing on the data mining and big data analytics methodologies. A federated or a hybrid approach leverages the strengths and mitigates the weaknesses of both data warehouse and big data analytics. We discuss the challenges, requirements, integrated models, components, scenarios and proposed solutions to the performance, efficiency, availability, security and privacy concerns in the context of smart cities. Our approach relies on several layers that run in parallel to collect and manage all collected data and create several scenarios that will be used to assist urban mobility. The data warehouse and big data analytics can serve as means to support clustering, classification, recommending systems, frequent item set mining. The challenge here is to populate the repository architecture with the schema, view definitions, metadata and specify/integrate the types of this architecture (Centralized Metadata repository, Distributed Metadata repository, Federated or Hybrid Metadata repository).

Keywords: ITS · Urban mobility · Big data analytics · Data mining · Data warehouse

1 Scope and Outline of the Paper

Plenty of solutions have been applied to mitigate and predict the number of road accidents. Like establishing stringent rules and regulations. But most of them failed to decrease road accidents. Many researchers work in the area of pervasive and context-aware computing has developed different kinds of context models. [4] approach contains different levels that represent the process of multimedia documents adaptation. [5] allows the production of medical, genetic and scientific data between health professionals, scientists and patients.

In [6] the author designs intelligent transport organization optimization schemes and analyzes the types of data processed on big data platform.

This paper attempts to integrate the servers, storage handling, knowledge management data and client tools for making decision. We provide an intelligent system for

© Springer Nature Switzerland AG 2020
M. Ezziyyani (Ed.): AI2SD 2019, LNNS 92, pp. 179–184, 2020.
https://doi.org/10.1007/978-3-030-33103-0_18

accident prevention and detection for human life safety. The prevention part involves the following aspects:

– Conduct surveys to analyze the driver behavior;
– Establish a decision support model for the detection and identification of accident factors;
– Analyze data and determine trends in road accidents and identify potentially dangerous accidents areas;
– Traffic flow forecasting;
– Emergency management;
– Urban mobility;
– Mitigating urban traffic congestion.

1.1   Big Data Analytics

The approach of Big data can support real-time access of metadata from source systems. It can also centrally and reliably maintain metadata definitions to the proper locations of the accurate definitions in order to improve performance and availability.

This architecture proposed in [3] present a brief summary of the various modules in Big data (Fig. 1).

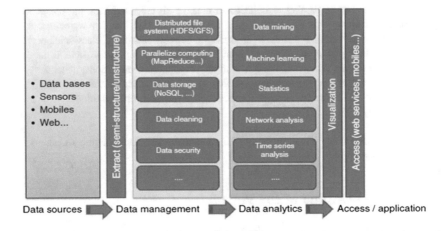

Fig. 1. Big data modules

Large volumes of data sets derived from sophisticated sensors and social media feeds are increasingly being used by the researchers. Processing a large amount of data is not easy with conventional parallel computing, due to the failure of the compute nodes and the scalability of the system. big data is a useful tool for improving the performance and availability.

The 4V's of big data – volume, velocity, variety and veracity makes the data management and analytics challenging for the traditional data warehouses.

Large data analysis - the process of analyzing and exploring large data - can generate operational and decisional information of a scale and specificity important for the control of urban mobility and energy efficiency. The need to analyze and exploit trend data collected by different sources is one of the main drivers of Big Data analytics tools.

Big Data Analytics is rapidly evolving both in terms of functionality and the underlying programming model. Such analytical functions support the integration of results derived in parallel across distributed pieces of one or more data sources [2].

1.2   Data Warehouse

This paper focuses on the use of data warehouse as a supporting tool in decision making in the context of smart cities. A data warehouse is a subject-oriented, integrated, time-variant and non-volatile collection of data in support of management's decision making process Inmon [1].

The data warehouse gathers database scattered from different sources. We focus on the benefits gained from using data warehouse.

2   Design and Architecture

The intelligent solution as it will be designed, will manage, analyze, alert, secure, and improve the quality of services offered by data management systems in data Warehouse, machine learning and Big Data analytics, these aspects allow the modeling of urban mobility scenarios, Traffic flow forecasting and auditing of computations and data.

2.1   Layer Acquisition, Cleaning, Loading Modeling Data

This view tries to provide a wide basis of integrated data and data modeling. These data come from several heterogeneous sources (Raw data, CSV, XML files, JSON, Sensors/embedded sensors, after the extracting phase, cleaning, filtering and the determination the schema, view definitions, metadata).

In this layer, we present a brief summary of the security and privacy aspects of our approach, the privacy requirements in urban mobility, and some of the existing privacy solutions in urban mobility.

The preservation of privacy largely relies on technological limitations on the ability to extract, analyze, and correlate potentially sensitive data sets. However, advances in Big Data analytics provide tools to extract and utilize this data, making violations of privacy easier. As a result, along with developing Big Data tools, it is necessary to create safeguards to prevent abuse (Bryant, Katz, & Lazowska, 2008).

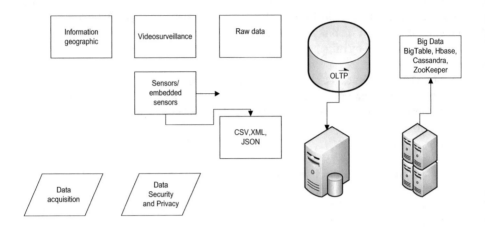

2.2 Layer Processing, Querying and Data Analysis

The major challenge of the road safety of this paper comes from the innovative character of the exploitation and the integration of some data associated to the field of road traffic management, the optimization of its infrastructure, as well as the trends, assets and constraints related to vehicles and pedestrians. Therefore, it is essential to exploit, in real time, these data and all the infrastructure that manages them.

We aim to design an integrated and scalable architecture to access a shared pool of configurable resources. This layer deals with the analysis, modeling and design of a system for access management and real-time detection of anomalies.

The major requirements is to preserve the integrity, confidentiality or availability of data in order to deduce recommendations on integrity of data-computations, and correctness-freshness and support for decision-making.

The need to handle querying and data analysis from various applications and data stores into the central repository may compromise data quality. We discuss how new technologies can improve urban mobility and contribute to road safety and congestion reduction in a smart city. Our approach is to gather, federate and synthesize data and support a decision support system consisting of a set of recommendations.

Data Mining Algorithm Applied to Intelligent Transport System: Development of data mining tools, and its integration into the part of the parallel processing of Big Data within the Big Data analytics management system and its coexistence with data warehouse.

2.3 Layer the Decision Support Technology

We discuss here how new technologies can improve urban mobility and contribute to road safety and congestion reduction in a smart city. Our approach is to gather, federate and synthesize data and support a decision support system consisting of a set of recommendations.

Modeling the ITS organization as a complex system: The ITS organization framework.

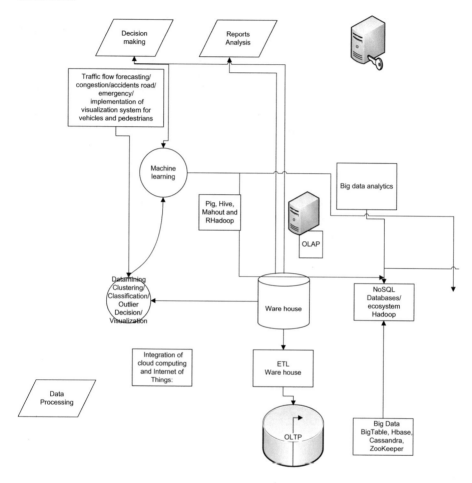

3 Conclusion

We have established how data analytics could benefit ITS using scenarios including efficient route guidance.

Regulations and policies of component organization in this architecture, and the general availability of the components are not discussed in detail. Our first challenge is the integration and conformity study of the different components of the proposed architecture.

References

1. Inmon, W.H.: Building the Data Warehouse, 2nd edn. Wiley, New York (1996)
2. El-Seoud, S.A., El-Sofany, H.F., Abdelfattah, M., Mohamed, R.: Big data and cloud computing: trends and challenges. IJIM **11**(2) (2017)
3. Ferandez, A., del Sara, R., López, V., Bawakid, A., del Jesus, M.J., Benitez, J.M., Herrera, F.: Big data with cloud computing: an insight on the computing environment, mapreduce, and programming frameworks. WIREs Data Min. Knowl. Discov. **4**, 380–409 (2014). https://doi.org/10.1002/widm.1134
4. Khallouki, H., Bahaj, M.: Context modeling architecture in pervasive computing environments for multimedia documents adaptation. In: 2016 5th International Conference on Multimedia Computing and Systems (ICMCS), pp. 611–615. IEEE (2016)
5. Abatal, A., Khallouki, H., Bahaj, M.: A semantic smart interconnected healthcare system using ontology and cloud computing. In: 2018 4th International Conference on Optimization and Applications (ICOA), pp. 1–5. IEEE, April 2018
6. Ying, C.: Intelligent transport decision analysis system based on big data mining. In: Advances in Computer Science Research (ACSR), vol. 73

MIMO-OFDM for Wireless Systems: An Overview

Mohamed Nassim Aarab$^{(\boxtimes)}$ and Otman Chakkor

National School of Applied Sciences, Tetouan, Morocco
nassimaarab0@gmail.com, otman.chakkor@uae.ac.ma

Abstract. Orthogonal frequency-division multiplexing (OFDM) effectively mitigates inter-symbol interference (ISI) caused by the delay spread of wireless channels. This paper describes the combination of MIMO system along with Orthogonal Frequency Division Multiplexing (OFDM) system which offers important features of both the system. Also, a comprehensive survey on OFDM for wireless communications. We address basic OFDM and related modulations, as well as techniques to improve the performance of OFDM for wireless communications, including channel estimation and signal detection, time- and frequency-offset estimation and correction, peak-to-average power ratio reduction, and multiple-input–multiple-output (MIMO) techniques.

Keywords: Multiple-input multiple-output (MIMO) · Inter-symbol interference · Orthogonal frequency division multiplexing (OFDM) · Space-time code · Peak to average power ratio (PAPR)

1 Introduction

Orthogonal frequency division multiplexing (OFDM) has become a popular technique for transmission of signals over wireless channels. Moreover MIMO-OFDM is the dominant air interface for 4G and 5G broadband wireless communications. It combines multiple-input, multiple-output (MIMO) technology, which multiplies capacity by transmitting different signals over multiple antennas, and orthogonal frequency-division multiplexing (OFDM), which divides a radio channel into a large number of closely spaced subchannels to provide more reliable communications at high speeds. Research conducted during the mid-1990s showed that while MIMO can be used with other popular air interfaces such as time-division multiple access (TDMA) and code-division multiple access (CDMA), the combination of MIMO and OFDM is most practical at higher data rates.

As OFDM suffers from strong intercarrier interference in this case. Therefore, in recent years an alternative to OFDM has been heavily promoted: Filter bank based multicarrier (FBMC) [1, 2].

Like OFDM, FBMC is also a multicarrier technique which employs per-subcarrier filtering. Filtering technique is used in FBMC. FMBC's improved synchronization and resistance to misalignments of frequency make the waveform an enticing alternative to OFDM [3]. However, the additional filtering required increases the implementation

© Springer Nature Switzerland AG 2020
M. Ezziyyani (Ed.): AI2SD 2019, LNNS 92, pp. 185–196, 2020.
https://doi.org/10.1007/978-3-030-33103-0_19

complexity. For short burst uplink communication with high filter length of FBMC is disadvantageous [4].

The applications which use advance wireless communication system require higher data rates, lower latency and efficient spectrum usage. The way to overcome the known limitations of OFDM is UFMC technique. The performance of UFMC signals in terms of side lobe attenuation is better than OFDM, BER for discrete narrow band networks [5]. Smart Gradient Project Active Constellation Extension (ACE-SGP), Tone Reservation (TR) methods are used to reduce the PAPR values for FBMC/OQAM signals [6]. UFMC does not have to use a cyclic prefix, although it can be used to improve the inter symbol interference (ISI) protection using special or unified structure of the frame [7].

This paper discusses several things, beginning first by describes the basic of MIMO-OFDM systems. The next section contains a brief introduction into Peak to average power ratio and some techniques to reduce it, Sect. 4 considers space–time coding techniques for MIMO-OFDM, while Sect. 5 discusses the performance of such a scheme with two receive antennas with and without channel estimation by Mat lab simulation, We then provide in Sect. 6 some of OFDM advantages & disadvantages, Sect. 7 wraps up with some open issues concluding remarks and perspective.

2   MIMO OFDM System

Multiple-In, Multiple-Out (MIMO) communication sends the same information as several signals simultaneously through multiple antennas, using a single radio channel. It uses multiple antennas to improve signal quality and strength of an RF link channel using antenna diversity. The data are divided into multiple data streams at the transmission point and rearranged on the receiver side by another MIMO radio configuration with the same number of antennas. While sending information via MIMO system can result in multipath propagation which leads as to inter-symbol interference when a wireless signal being transmitted reaches a receiver through different paths. This commonly occurs when reflected signals bounce off of surfaces, when the wireless signal refracts through obstacles, and because of atmospheric conditions. These paths have different lengths before reaching the receiver, thus creating different versions that reach at different time intervals. The delay in symbol transmission will interfere with correct symbol detection. The amplitude and or phase of the signal can be distorted when the different paths are received for additional interference (Fig. 1).

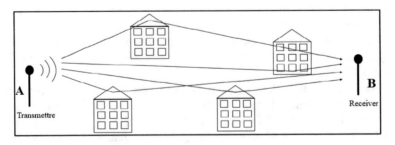

Fig. 1. Multipath propagation effect

Inter-symbol interference is a signal distortion in telecommunication systems. One or more symbols can interfere with other symbols causing noise or a less reliable signal. This has the effect of a blur or mixture of symbols, which can reduce signal clarity. If inter-symbol interference occurs within a system, the receiver output becomes erroneous at the decision device. This is an undesired result that should be reduced to the most minimal amount possible. Error rates from inter-symbol interference are minimized through the use of adaptive equalization techniques and error correcting codes.

The MIMO technology can be used with full potential to maximize the multi-path channel between the transmitter and receiver, and transmit parallel multi-channel data independently. This leads to enhance the spectrum efficiency. MIMO can resist multi-path fading, however for frequency selective deep fading, the MIMO system remains powerless. The solution of this complexity is to use MIMO system along with OFDM which is the core technology of the next-generation mobile communications. OFDM has distinctive feature that it has unique frequency domain block modulation and frequency domain channel equalization technology [8].

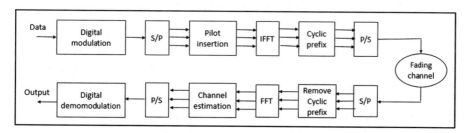

Fig. 2. Simplified block diagram of MIMO-OFDM system

OFDM is essentially a discrete implementation of multicarrier modulation, which divides the transmitted bit-stream into many different sub-streams and sends them over many different subchannels. Typically, the subchannels are orthogonal and the number of subchannels are chosen such that each subchannel has a bandwidth much less than the coherence bandwidth of the channel. This, inter-symbol interference (ISI) on each subchannel will be very small. For this reason, OFDM is widely used in many high data rate wireless systems.

Figure 2 shows a simplified block diagram of an N-tone OFDM system. First, the incoming bits are mapped to data symbols according to some modulation scheme such as QPSK or QAM. Then the serial data stream is converted into a number of parallel blocks, and each of them has length-N. Then, each block of symbols (including pilot symbols, which are used for channel estimation or synchronization) will be forwarded to the IFFT and transformed into an OFDM signal. After that, the OFDM signal will be appended with a cyclic prefix (CP) by copying the last Ncp samples to the top of the current OFDM block. By choosing the length of the cyclic prefix larger than the maximum path delay of the channel, ISI can be eliminated [9]. Afterward, the OFDM blocks will be converted to serial signals and sent out.

At the receiver, assuming a perfect timing and carrier frequency synchronization, the received signals will be first converted to parallel signals and then the cyclic prefix (CP) will be removed. After going through the DFT block, the data symbols are detected with the estimated channel information. After demodulation, the transmitted bit stream is recovered.

3 Peak to Average Power Ratio

The PAPR is the relation between the maximum power of a sample in a given OFDM transmit symbol divided by the average power of that OFDM symbol. PAPR occurs when in a multicarrier system the different sub-carriers are out of phase with each other [10]. At each instant they are different with respect to each other at different phase values. When all the points achieve the maximum value simultaneously; this will cause the output envelope to suddenly shoot up which causes a 'peak' in the output envelope. Due to presence of large number of independently modulated subcarriers in an OFDM system, the peak value of the system can be very high as compared to the average of the whole system. This ratio of the peak to average power value is termed as Peak-to Average Power Ratio.

An OFDM signal consists of a number of independently modulated sub-carriers which can give a large PAPR when added up coherently. When N signals are added with the same phase, they produce a peak power that is N times the average power of the signal. So OFDM signal has a very large PAPR, which is very sensitive to nonlinearity of the high-power amplifier. In OFDM, a block of N symbols $\{X, k = 0, 1. \ldots, N-1\}$ k, is formed with each symbol modulating one of a set of subcarriers, $\{f, k = 0, 1. \ldots, N-1\}$ k. The N subcarriers are chosen to be orthogonal, that is, $f_k = D$, where $D_f = 1\ N_T$ and T is the original time period. The resulting signal is given as:

$$x(t) = \sum_{n=0}^{N-1} X_k e^{j2\pi f_k t} \qquad 0 \le t \le N_T \tag{1}$$

PAPR is defined as:

$$PARP = \left(\frac{\max\left\{ |x[t]|^2 \right\}}{E\left\{ |x[t]|^2 \right\}} \right) \tag{2}$$

Where, $|x[t]|$ is the amplitude of $x[t]$ and E denote the expectation of the signal.
PAPR in dB can be written as:

$$PARP(dB) = 10 \log_{10}(PARP) \tag{3}$$

PAs at the transmitter are driven into saturation due to high PAPR [11, 12], degrading the BER performance. To avoid driving the PA into saturation, the average power of the signal may be reduced. However, this reduces the SNR and, consequently, the BER performance. Therefore, it is preferable to solve the problem of high PAPR by reducing the peak power of the signal. Many PAPR reduction techniques have been proposed. The performance of a PAPR reduction scheme is usually demonstrated by two main factors: the complementary cumulative distributive function (CCDF) and bit error rate (BER).

3.1 PAPR Reduction Techniques

- Clipping and Filtering

One of the simplest PAPR reduction methods is the method of clipping the high peaks of the OFDM signal before passing it through the power amplifier. Clipping is a nonlinear process that leads to both in-band and out-of-band distortions. While in-band distortion causes spectral spreading and can be eliminated by filtering the signal after clipping, Out-of-band distortion can degrade the BER performance and cannot be reduced by filtering. However, oversampling by taking longer IFFT can reduce the in-band distortion effect as portion of the noise is reshaped outside of the signal band that can be removed later by filtering. Filtering the clipped OFDM signal can preserve the spectral efficiency by eliminating the out of band distortion and, hence, improving the BER performance but it can lead to peak power regrowth [14].

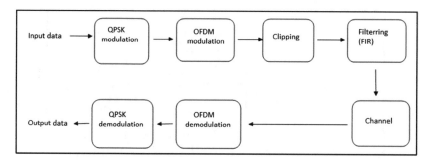

Fig. 3. Block diagram of OFDM with clipping and filtering

A block diagram of clipping and filtering is shown in Fig. 2 [14]. Quadrature phase shift-keying (QPSK) data symbols are passed through an inverse fast Fourier transform (IFFT) module to realize the OFDM modulation. If the digital OFDM signals are clipped directly, the resulting clipping noise will all fall in-band and cannot be reduced by filtering. Hence an oversampling is done. Then, the real-valued bandpass samples, x, are clipped at an amplitude A as follows:

$$y = \begin{cases} -A, & \text{if } x < -A \\ -x, & \text{if } -A \leq x \leq A \\ A, & \text{if } x < A \end{cases} \tag{4}$$

Then the clipped signal is passed through a filter and transmitted. At the receiver, the reverse operations are done.

- Selective Mapping

Selective mapping (SLM) is a simple approach to reduce PAPR [15]. In this method, a set of sufficient different OFDM symbols $x_m, 0 \leq m \leq M - 1$ are generated, each of length N, all representing the same information as that of the original OFDM symbol x, then the one with the least PAPR is transmitted. Mathematically, the transmitted OFDM symbol x is represented as

$$\hat{x} = argmin_{0 \leq m \leq M-1} [PAPR(x_m)] \qquad (5)$$

The OFDM symbols set can be generated by multiplying the original data block $X = [X_1, X_2, \ldots \ldots X_N]$, element-by-element, by M different phase sequences pm, each of length N, before taking IDFT. For simplifying the implementation, the phase sequences p_m can be set to $\{\pm 1, \pm i\}$ as these values can be implemented without multiplication. Then the modified OFDM symbol $x_m, 0 \leq m \leq M - 1$, is the IDFT of the element-by element multiplication of X and p_m (Fig. 4).

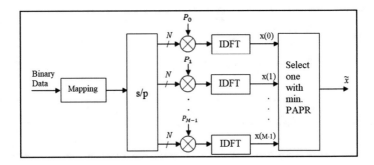

Fig. 4. Block diagram of OFDM transmitter with SLM [13]

$$x_m = IDFT\left[X_1 e^{j\varphi m,1} X_2 e^{j\varphi m,2} \ldots X_N e^{j\varphi m,N}\right] \qquad (6)$$

A block diagram of the SLM technique is depicted in Fig. 3 [10]. The selected phase sequence should be transmitted to the receiver as side information to allow the recovery of original symbol sequence at the receiver, which decreases the data transmission rate. The phase sequences $p_m, 0 \leq m \leq M - 1$ should be stored at both the transmitter and receiver. An erroneous detection of the side information will cause the whole system to be destroyed. Hence strong protection of the side information is very important.

- Partial Transmit Sequence

In partial transmit sequence (PTS), an input data block having length N is partitioned into a number of disjoint sub-blocks by pseudo-random partitioning [16]. The IDFT is computed separately for each one of these sub-blocks and then weighted by a phase factor. The phase factors are selected such that the PAPR of the combined signal of all the sub-blocks is minimized. Figure 5 shows a block diagram of the OFDM transmitter with PTS technique [12]. Let an input data block X be partitioned into M disjoint sub-blocks, $X_m = [X_{m,1}, X_{m,2}, \ldots X_{m,N}]$, $1 \leq m \leq M$ such that any two of these sub-blocks are orthogonal and X is the combination of all the M sub-blocks.

$$X = \sum_{M=1}^{M} X_m \tag{7}$$

Then the IDFT for each sub-block, $x_m, 1 \leq m \leq M$, is computed and weighted by a phase factor $b_m = e^{j\varphi m}$, $\varphi_{m,k} \in [0, 2\pi) 1 \leq m \leq M$. The aim now is to select the set of phase factors b_m's that minimizes the PAPR of the combined time domain signal x, where x is defined as

$$x = \sum_{M=1}^{M} x_m b_m \tag{8}$$

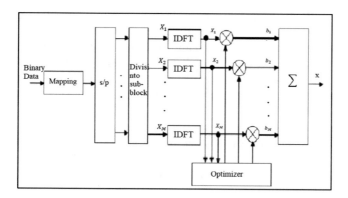

Fig. 5. Block diagram of OFDM transmitter with PTS [13]

The first phase factor, b_1 can be set to 1 in order to avoid any loss of performance, therefore, M-1 phase factors are to be found by an exhaustive search. On one hand, the larger is the number of sub-blocks M, the greater is the reduction in PAPR. On the other hand, the search complexity is increasing exponentially with M. In addition, M IDFT blocks are needed to implement the PTS scheme, side information has to be transmitted.

- Linear Block Coding

Instead of dedicating some bits of the codeword to enhance BER performance, these bits are now dedicated to reduce PAPR [11]. The codewords with low PAPR have to be chosen for transmission. A simple linear block coding (LBC) scheme was proposed in [17], where 4 bits are mapped into 5 bits by adding a parity bit. It is based on the observation that irrespective of codeword length, four specific codewords will always have disproportionately large PAPR values [18]. These are the codewords where the odd and even bit values are equal, i.e. the all-zero, all-one, (1010 …) and (0101 …) codewords. The PAPR can hence be very easily reduced by eliminating these codewords using a simple added bit code. If the codeword length is equal to n, then a single extra bit b_{n+1}, is added with a value equal to the inverse of the penultimate codeword bit. Then the four codewords with high power are now eliminated.

4   MIMO Space-Time Block Coding and Alamouti Codes

In order that MIMO spatial multiplexing can be utilised, it is necessary to add coding to the different channels so that the receiver can detect the correct data.

4.1   Space Time Block Codes

Space-Time Codes are a powerful scheme that combines coding with transmit diversity to achieve high diversity performance in wireless systems. They are used for MIMO systems to enable the transmission of multiple copies of a data stream across a number of antennas and to exploit the various received versions of the data to improve the reliability of data-transfer. Space-time coding combines all the copies of the received signal in an optimal way to extract as much information from each of them as possible.

Space time block coding uses both spatial and temporal diversity, in this way enables significant gains to be made.

Space-time coding involves the transmission of multiple copies of the data. This helps to compensate for the channel problems such as fading and thermal noise. Although there is redundancy in the data some copies may arrive less corrupted at the receiver.

When using space-time block coding, the data stream is encoded in blocks prior to transmission. These data blocks are then distributed among the multiple antennas (which are spaced apart to decorrelate the transmission paths) and the data is also spaced across time.

A space time block code is usually represented by a matrix. Each row represents a time slot and each column represents one antenna's transmissions over time.

$$\begin{bmatrix} S_{11} & \cdots & S_{1n} \\ \vdots & \ddots & \vdots \\ S_{m1} & \cdots & S_{mn} \end{bmatrix} \tag{9}$$

Within this matrix, Sij is the modulated symbol to be transmitted in time slot i from antenna j. There are to be T time slots and n_T transmit antennas as well as n_R receive antennas. This block is usually considered to be of 'length' T.

4.2 MIMO Alamouti Coding

A particularly elegant scheme for MIMO coding was developed by Alamouti. The associated codes are often called MIMO Alamouti codes or just Alamouti codes.

The MIMO Alamouti scheme is an ingenious transmit diversity scheme for two transmit antennas that does not require transmit channel knowledge.

The Alamouti code encoder as illustrated in Fig. 3 where the information bits are first modulated using an M-ary modulation scheme. The encoder then takes a block of two modulated symbols X_1 and X_2 in each encoding operation and passes it to the transmit antennas according to the code matrix,

$$X = \begin{pmatrix} X_1 & -X_2^* \\ X_2 & X_1^* \end{pmatrix}$$ (10)

The first and second column of (10) represent the first and second transmission period, respectively. The first row corresponds to the symbols transmitted from the first antenna and the second row corresponds to the symbols transmitted from the second antenna. Furthermore, during the first and second symbol periods, both the first and second antenna transmits the signal. This implies that the transmission process occurs both in space (across two antennas) and time (two transmission periods) [19] (Fig. 6).

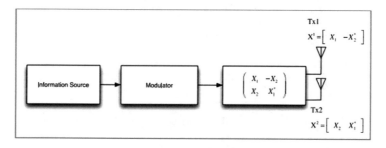

Fig. 6. Block diagram of the Alamouti space-time encoder

5 Simulation

In this section, we study the performance of 2 × 1 and 2 × 2 MIMO OFDM systems using QPSK modulation technique, the simulation is done using MATLAB simulation software.

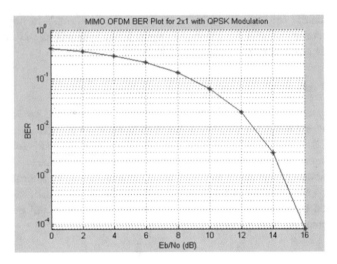

Fig. 7. MIMO OFDM BER plot for 2 × 1 with QPSK modulation

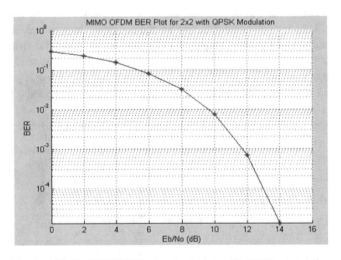

Fig. 8. MIMO OFDM BER plot for 2 × 2 with QPSK modulation

The simulated result in Figs. 7 and 8 for the MIMO OFDM system shows a plot of BER v/s SNR for 2 × 1 and 2 × 2 systems with QPSK modulation. The figure indicates that at BER of 10^{-3} point the performance in SNR is ~ 15 dB and 12 dB in 2 × 1 and 2 × 2 systems respectively. The figure also shows that at 10^{-3} BER, when compared to 2 × 1, there is an improvement of SNR ~ 3 dB in 2 × 2 MIMO systems.

6   OFDM Advantages and Disadvantages

- **OFDM advantages:** OFDM has been used in many high data rate wireless systems because of the many advantages it provides.
  - Immunity to selective fading: One of the main advantages of OFDM is that is more resistant to frequency selective fading than single carrier systems because it divides the overall channel into multiple narrowband signals that are affected individually as flat fading sub-channels.
  - Resilience to interference: Interference appearing on a channel may be bandwidth limited and, in this way will not affect all the sub-channels. This means that not all the data is lost.
  - Resilient to ISI: Another advantage of OFDM is that it is very resilient to intersymbol and inter-frame interference. This results from the low data rate on each of the sub-channels.
  - Simpler channel equalization: One of the issues with CDMA systems was the complexity of the channel equalization which had to be applied across the whole channel. An advantage of OFDM is that using multiple sub-channels, the channel equalization becomes much simpler.
- **OFDM disadvantages:** Whilst OFDM has been widely used, there are still a few disadvantages to its use which need to be addressed when considering its use.
  - High peak to average power ratio: An OFDM signal has a noise like amplitude variation and has a relatively high large dynamic range, or peak to average power ratio. This impacts the RF amplifier efficiency as the amplifiers need to be linear and accommodate the large amplitude variations and these factors mean the amplifier cannot operate with a high efficiency level.
  - Sensitive to carrier offset and drift: Another disadvantage of OFDM is that is sensitive to carrier frequency offset and drift. Single carrier systems are less sensitive.

7   Conclusion

MIMO-OFDM is an attractive transmission technique for high bit rate communication systems. MIMO-OFDM has become a promising technology for high performance to broadband wireless communications, we were interested in this paper with the major features of MIMO links for use in future wireless networks. We start with the basic principle of OFDM and techniques to deal with impairments in wireless systems, including peak to average power ratio, channel estimation, Space-Time Code. Then, we study the performance of such a scheme with two receiver antennas with and without channel estimation using MATLAB.

As we march towards 2020, research on MIMO will continue to mature, and will develop to mmWave massive MIMO, and new trends will emerge. Though there are many challenges to address, mmWave massive MIMO shows amazing prospects and potentials in realizing the 1000-fold capacity quest for 5G cellular networks. The technology will, no doubt, usher in new paradigms for next-generation mobile networks and open up new frontiers in cellular services and applications.

References

1. Farhang-Boroujeny, B.: OFDM versus filter bank multicarrier. IEEE Sig. Process. Mag. **28**(3), 92–112 (2011)
2. Bellanger, M.: Physical layer for future broadband radio systems. In: 2010 IEEE Radio and Wireless Symposium (RWS), pp. 436–439, 10–14 January 2010
3. Kibria, M.G., Villardi, G.P.: Coexistence of Systems with Different Multicarrier Waveforms in LSA Communications arXiv:1708.04431 [cs.IT], 15 August 2018
4. Wunder, G., Jung, P., Kasparick, M., Schaich, F., Chen, Y., Brink, S., Gasper, I., Michailow, N., Festag, A., Mendes, L., Cassiau, N., Ktenas, D., Dryjanski, M., Picerzykr, S., Eged, B., Vago, P., Wiedmann, F.: 5GNOW: nonorthogonal, asynchronous waveforms for future mobile applications. IEEE Commun. Mag. **52**, 97–105 (2014)
5. Geng, S., Xiong, X., Cheng, L., Zhao, X., Huang, B.: UFMC system performance analysis for discrete narrowband private networks. In: 2015 IEEE 6th International Symposium Microwave, Antenna, Propagation, and EMC Technologies (MAPE), 14 July 2016
6. Laabidi, M., Zayani, R., Roviras, D., Bouallegue, R.: PAPR reduction in FBMC/OQAM systems using active constellation extension and tone reservation approaches. In: IEEE Symposium on Computers and Communication (ISCC), pp. 657–662, 6–9 July 2015
7. Schaich, F., Wild, T., Chen, Y.: 5G air interface design based on universal filtered (UF-) OFDM. In: Proceedings of 19th International Conference on Digital Signal Processing, pp. 699–704 (2014)
8. Gupta, B., Saini, D.S.: BER analysis of space frequency block coded MIMO-OFDM systems using different equalizers in quasi-static mobile radio channel. In: IEEE-CSNT, pp. 520–524 (2011)
9. Li, Y., Stüber, G.L.: Orthogonal Frequency Division Multiplexing for Wireless Communications. Springer, Heidelberg (2007)
10. Gangwar, A., Bhardwaj, M.: An overview: peak to average power ratio in OFDM system & its effect. Int. J. Commun. Comput. Technol. **1**(2), 22–25 (2012)
11. Rahmatallah, Y., Mohan, S.: Peak-to-average power ratio reduction in OFDM schemes: a survey and taxonomy. IEEE Commun. Surv. Tutorials **15**(4), 1567–1592 (2013)
12. Han, S.H., Lee, J.H.: An overview of peak-to-average power ratio reduction techniques for multicarrier transmission. IEEE Wirel. Commun. **12**(2), 56–65 (2005)
13. Ann, P.P., Jose, R.: Comparison of PAPR reduction techniques in OFDM systems. In: International Conference on Communication and Electronics Systems (ICCES) (2016)
14. Li, X., Cimini,Jr., L.J.: Effect of clipping and filtering on the performance of OFDM. IEEE Commun. Lett. **2**(5), 71–75 (1998)
15. Buml, R., Fischer, R., Huber, J.: Reducing the peak-to-average power ratio of multicarrier modulation by selected mapping. Electron. Lett. **32**(22), 2056–2057 (1996)
16. Muller, S.H., Huber, J.B.: OFDM with reduced peak-toaverage power ratio by optimum combination of partial transmit sequences. IEE Electron. Lett. **33**(5), 368–369 (1997)
17. Fragiacomo, S., Matrakidis, C., O'Reilly, J.J.: Multicarrier tranismission peak-to-average power reduction using simple block code. Electron. Lett. **34**(14), 953–954 (1998)
18. Jones, A.E., Wilkinson, T.A., Barton, S.K.: Block coding scheme for reduction of peak to mean envelope power ratio of multicarrier transmission schemes IEEE Electron. Lett. **30**(22), 2098–2099 (1994)
19. Asim, M., Crosby, G.V.: Reducing the peak to average power ratio of MIMO-OFDM systems, p. 40 (2008)
20. Naguib, A.F., Tarokh, V., Seshadri, N., Calderbank, A.R.: Space-time codes for high data rate wireless communication: mismatch analysis. In: Proceedings of IEEE International Conference on Communications, pp. 309–313, June 1997

Optimization and Evolutionary Games, Stochastic Equilibrium Application to Cellular Systems

Sara Riahi[1]([✉]) and Azzeddine Riahi[2]

[1] Department of Computer Science, Faculty of Sciences,
Chouaib Doukkali University, El Jadida, Morocco
riahisaraphd@gmail.com
[2] IMC Laboratory, Faculty of Sciences, Chouaib Doukkali University,
El Jadida, Morocco
riahikh@gmail.com

Abstract. LTE systems are designed to serve different classes of traffic through the IP-based packet-switched networks. Because of the inconsistent QoS requirements for each traffic class, LTE systems have scheduling mechanisms to support service differentiation when allocating block resources. As the 3GPP standard does not require the adoption of a particular approach, the schedulers design is left open to researchers and designers. This work focuses first on a study of some research that has addressed the management of resources in LTE networks. The study presents a classification of schedulers in the uplink and is interested in the class of schedulers based QoS because of the importance of delay parameters and flow in optimizing the management of resources. Then, some scheduling algorithms in the downlink are exposed in order to make a complete analysis of the different aspects adopted in the scheduling. Secondly, the resource optimization algorithm in the uplink in fixed WIMAX networks is presented. The algorithm defines a priority management policy to improve the low priority traffic service without affecting the high priority traffic QoS. Finally, an evaluation of existing solutions is carried out to a design of a robust scheduling mechanism.

Keywords: LTE · Resource allocation · Schedulers · Uplink · Equilibrium · QoS

1 Introduction

For all operators and for many years, voice calls were among the sources of income, the demand for mobile data has been initially slow, but from 2010 and through the discovery of smart phones, customers have become quite demanding, they recommend new areas with better quality of service (QoS). LTE (Long Term Evolution) has emerged as a promising solution to meet customer requirements. The implementation of LTE is being developed in several countries around the world including Morocco belongs. A key feature of the LTE network is the adoption of a number of advanced radio resource management procedures to improve system performance and to meet the

© Springer Nature Switzerland AG 2020
M. Ezziyani (Ed.): AI2SD 2019, LNNS 92, pp. 197–206, 2020.
https://doi.org/10.1007/978-3-030-33103-0_20

needs of users in terms of quality of service especially for real-time streams namely voice and video [1]. In particular, packet scheduling mechanisms play a fundamental role because they are responsible for choosing carefully and for a specific time, how to distribute the radio resources between the different stations taking into account channel conditions and QoS requirements. The 3GPP standards have not yet specified the access network radio resource allocation algorithm. This is a search path for developers and institutions [2] (Fig. 1).

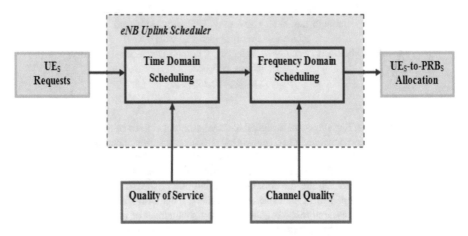

Fig. 1. Packet scheduling overview in LTE

The main contribution of this work is spread over two parts, the first is to prove which of the algorithms, the best known in mobile radio networks, warrants good performance for each type of real-time flows, especially for VoIP and Video stream, and for extreme scenarios, while for the second part, the objective is to develop and evaluate the performance of a new scheduling algorithm, we compare the performance with that of well known algorithms, already implemented in the eNodeB base stations and were chosen as standards with their performances in mobile radio networks [3]. Our performance evaluation is conducted by considering different cases of study scenarios, in a simulation environment, using the LTE-Sim simulator. The selected performance criteria are the packet transmission delay, packet loss rate (PLR), packet throughput and spectral efficiency.

2   Scheduling Resources in LTE

The most important goal in LTE is the scheduling of radio resources to meet the QoS requirements of all applications. This objective is very difficult to achieve in the presence of real-time multimedia applications with intolerance packet transmission delay. In LTE systems, it is exploited that users do not have the same sensitivity to disturbances of the mobile radio channel [4]. Indeed, in the presence of multiple users, it is always possible to find a user with a good quality of the signal transmitted on the radio channel. Based on this idea, MT (Maximum Throughput) and PF (Proportional Fair) scheduling algorithms have become the most well-known algorithms in scheduling strategies in LTE networks, because allocation decisions are strictly based on the quality of the channel experienced by each UE that periodically measures the quality of the channel from the reference symbols and informs the eNB through Channel Quality Information (CQI) messages. This information is subsequently used by the scheduler to distribute the RBs (Resource Block) between the users.

This information is also used by the AMC (Adaptive Modulation Coding) module which selects for each UE the appropriate modulation and coding scheme to maximize spectral efficiency. LTE-Sim implemented the scheduling classes in direction UL and DL and called ULScheduler and DLScheduler [5]. These classes provide the implementation of some of the most scheduling algorithms commonly encountered in scheduling strategies: At the beginning of each subframe, the scheduler selects all the streams that can be scheduled. A stream can be scheduled if and only if it contains packet data to be transmitted in the MAC layer and the receiver to which the stream is destined is in an active state. At each TTI, the scheduler calculates a metric for each flow that can be scheduled. We assume that the metric assigned to stream i on subchannel j is defined by $w_{i,j}$. Different algorithms according to the method of calculating the metric [6]. The scheduler operates by assigning the subchannel j to the stream i having the largest metric (Fig. 2).

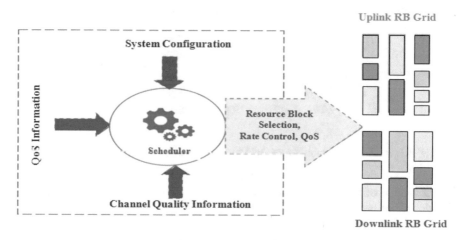

Fig. 2. LTE eNODE scheduler

In LTE, the scheduling procedure can be summarized by the following steps [7]:

- The eNB creates a list of DL-sense streams having data packets to transmit, that is, the streams that can be scheduled for the current sub-frame.
- The channel quality parameters (CQI) and the length of the MAC layer queues are recorded for each stream.

- According to the scheduling strategy, the metric is calculated for each selected stream from the previous list [8].
- The eNB assigns each subchannel flow having the maximum metric on the subchannel. It is important to note that during the allocation process, the eNB takes into account the amount of data already transmitted by each flow. Therefore, when the packets constituting a stream are emptied from the queue, this stream is immediately removed from the list of flows whose scheduling is possible.
- For each scheduled flow, the eNB calculates the size of the data block to be transmitted, that is, the amount of data that will be transmitted at the MAC layer during a TTI. The eNB uses the AMC module to match the channel quality (CQI) of each stream to the appropriate modulation and coding scheme (Fig. 3).

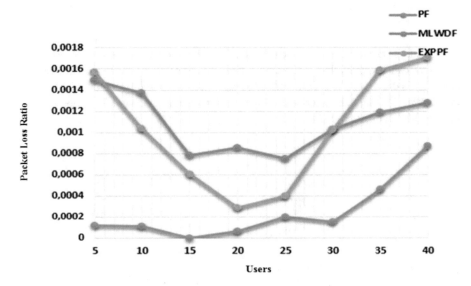

Fig. 3. Variation in the rate of loss of VoIP packets according to the number of users

At the end of the allocation procedure, the eNB of all scheduled flows to estimate the number of packets remaining in the queues of the MAC layer [9].

In order to obtain the metric, the schedulers generally need to know the average transmission rate $\overline{R}_l$ of the stream i, and the instantaneous rate available for the UE on the subchannel j. This approach is important when the metric takes into account the past performance of the scheduled flows in order to balance the distribution of resources between the UEs. In particular, at each TTI, the estimate of $\overline{R}_l$ is given by [10]:

$$\overline{R}_l(k) = 0.8\overline{R}_l(k-1) + 0.2R_i(k) \tag{1}$$

Where $R_i(k)$ the bit rate is allocated to the i flow during the kth TTI and $\overline{R}_l(k-1)$ is the estimate of the average transmission rate at $(k-1)^{\text{éme}}$ TTI.

From the following, here is a description of the implemented scheduling algorithms.

3   The PF Algorithm

This scheduler combines the radio resources by jointly taking into account the quality of the channel and the throughput completed by each user in the previous TTIs [11]. The goal of this algorithm is to maximize the total throughput of the network and to ensure fairness between flows. For this algorithm, the metric $w_{i,j}$ is defined as the ratio between the instantaneous flow rate available for the $i(r_{i,j})$ streams and the average flow rate (calculated at time $k-1$) of the flow i.

$$w_{i,j} = \frac{r_{i,j}}{\overline{R}_l} \tag{2}$$

The value of $r_{i,j}$ is calculated by the AMC module by considering the value of the CQI on the subchannel j sent by the UE to which the stream i is intended.

4   The M-LWDF Algorithm

This scheduler supports a user who can request multiple services with different QoS requirements. For each real-time stream, considering the maximum delay τ_i, the probability δ_i is defined as the maximum probability that the delay of the first packet of the queue $D_{HOL,i}$ exceeds the maximum fixed delay.

In order to give priority to the real time streams having the greatest delay (delay of the first packet of the queue) and having the best propagation conditions on the radio channel, the metric has been defined in this scheduler by [12]:

$$w_{i,j} = \alpha_i D_{HOL,i} \frac{r_{i,j}}{\overline{R}_l} \tag{3}$$

Where $r_{i,j}$ and $\overline{R}_l$ have the same meanings as in the preceding paragraphs and α_i is given by:

$$\alpha_i = -\frac{log\delta_i}{\tau_i} \tag{4}$$

For non-real time traffic or nRTPS, metric is calculated from the equation provided by the PF algorithm.

In the current version of LTE-Sim, the M-LWDF algorithm is implemented in such a way that packets belonging to a real-time stream are removed from the MAC layer queue if they are not transmitted before expiration of the maximum wait time. This is necessary to avoid waste of bandwidth. This implementation is not implemented in the PF algorithm since it is not designed for real-time services [13].

5 The EXP PF Algorithm

This scheduler is designed to improve the priority of real-time stream while allowing non-real time traffic to benefit from a minimum of satisfaction. The queue header packet of these streams has a delay very close to the maximum delay. For real-time streams, the metric is calculated using the following equations [14]:

$$w_{i,j} = exp\left(\frac{\alpha_i D_{HOL,i} - \chi}{1 + \sqrt{\chi}}\right)\frac{r_{i,j}}{\overline{R_l}} \qquad (5)$$

Where the symbols have the same meanings as before and the value of χ is given by [15]:

$$\chi = \frac{1}{N_{rt}}\sum_{i=1}^{N_{rt}}\alpha_i D_{HOL,i} \qquad (6)$$

With N_{rt} the number of active real-time flows in the direction DL. Moreover, the metric of non-real time traffic are calculated using the metric of the PF algorithm. In the EXP PF algorithm, real-time packets whose timeout to exceed the maximum delay before transmission are removed from the queue in the MAC layer [16].

6 Simulation and Implementation of Schedulers

Scheduling is one of the most important resource management mechanisms in LTE networks, it determines which user should be transmitted in a given time interval (Fig. 4).

This is a critical design element since it allocates the allocation of the channel among users and thus, in general, determines the overall behavior of the system. An optimal system throughput can be achieved by assigning all radio resources to the user with the best channel radio conditions; however a Scheduling in practice should have several levels of fairness.

Thus, by choosing different scheduling algorithms, the operators can adapt the behavior of the system to their needs. So it is not necessary to standardize the algorithms used, instead, operators can choose different criteria. Prediction of channel quality, cell capacity, and different classes of traffic priorities are examples of information on which the Scheduler could base decisions.

The simulation of a complete PHY layer is not compatible with complex scenarios, in fact, the simulation of a complete PHY transmission between two nodes in an LTE network can take hours. This difficulty was circumvented in LTE-Sim adopted a model based on an analytical approach. This approach was adopted the same with success in the following simulations (Fig. 5).

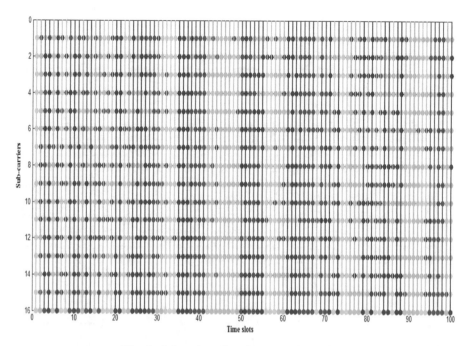

Fig. 4. Subcarriers allocation according to PFS

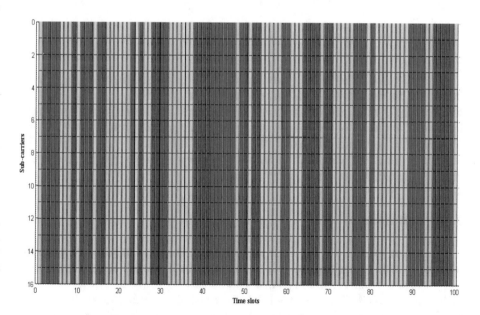

Fig. 5. Subcarriers allocation according to M-LWDF

From simulations, we found that non-real-time data traffic strongly influences the behavior of scheduling algorithms. It is clear that PF behaves badly for real-time traffic, it does not respect the time constraint, and the rate of loss of packets is important. In addition, the Proportional Fair which provides the best throughput for different traffic types (Fig. 6).

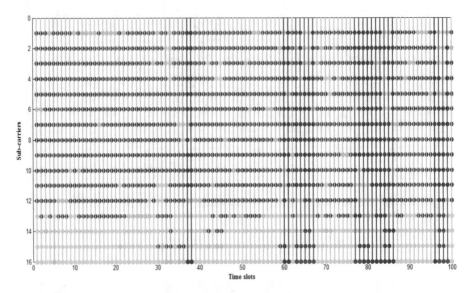

Fig. 6. Subcarriers allocation according to EXPPF

MLWDF and EXPPF allow guaranteeing a quality of service for the real time traffic in favor of the throughput. If the network load in real-time traffic is not important, we find that the PF is the most robust scheduling mechanism. If the load of real-time traffic increases, the quality of service offered by the PF deteriorates considerably and it is recommended to adopt the MLWDF or the EXPPF. Finally the choice of the scheduling policy will depend on several factors mainly the type of traffic, the load of the network and the number of users to serve.

7 Conclusion

The allocation algorithms have been implemented in order to measure the impact of different designs on performance indicators and to check whether the new approaches can improve scheduling in the uplink without degrading the quality of service of the high priority class. LTE technology has been standardized by 3GPP, it is based on an all-IP architecture that allows providing and guaranteeing the broadband for the users with a better management of quality of service between the different types of service and the different categories of the users. In our work, we presented the LTE architecture as well as the definition of quality of service on the LTE domain with all the parameters that act on the differentiation between the IP flows. In the simulation part, we focused our work on the study and the evaluation of the performances of the different scheduling algorithms (PF, MLWDF, EXPPF) proposed for the LTE technology. From the results obtained at the end of the simulation, it was found that the choice of the most compliant scheduler depends on the type of traffic expected, the network load (estimated user number) and the coverage.

Acknowledgment. We would like to thank the CNRST of Morocco (I 012/004) for support.

References

1. Riahi, S., Riahi, A.: Game theory for resource sharing in large distributed systems. Int. J. Electr. Comput. Eng. IJECE **9**(2), 1249–1257 (2019). https://doi.org/10.11591/ijece.v9i2. ISSN: 2088-8708
2. Riahi, S., Riahi, A.: Energy Efficiency analysis in wireless systems by game theory. In: The 5th International IEEE Congress on Information Sciences and Technology, IEEE Smart Cities and Innovative Systems in Marrakech, Morocco, 21–24 October 2018
3. Riahi, S., Riahi, A.: Optimal performance of the opportunistic scheduling in new generation mobile systems. In: International Conference on Smart Digital Environment (ICSDE 2018), Rabat-Morocco, 18–20 October 2018
4. Riahi, S., Riahi, A.: Resource allocation optimization based on channel quality for long term evolution systems (LTE). J. Theoret. Appl. Inf. Technol. **97**(6) (2019). ISSN: 1992-8645
5. Riahi, S., Elhore, A.: Estimation of QoS in 4th generation wireless systems. In: The International Conference on Learning and Optimization Algorithms: Theory and Applications (LOPAL 2018), Rabat, Morocco, 2–5 May 2018
6. Baiocchi, A., Chiaraviglio, L., Cuomo, F., Salvatore, V.: Joint management of energy consumption maintenance costs and user revenues in cellular networks with sleep modes. IEEE Trans. Green Commun. Netw. **1**, 167–181 (2017). ISSN 2473-2400
7. Yuan, G.X., Zhang, X., Wang, W.B., Yang, Y.: Carrier aggregation for LTE-advanced mobile communication systems. IEEE Commun. Mag. **48**, 88–93 (2010)
8. Vatsikas, S., Armour, S., De Vos, M., Lewis, T.: A distributed algorithm for wireless resource allocation using coalitions and the nash bargaining solution. In: IEEE Vehicular Technology Conference (VTC), pp. 1–5, May 2011
9. Toskala, A., Holma, H., Pajukoski, K., Tiirola, E.: UTRAN long term evolution in 3GPP. In: Proceedings of IEEE Personal Indoor and Mobile Radio Communications Conference (PIMRC 2006), September 2006
10. Wang, C., Huang, Y.-C.: Delay-scheduler coupled throughput-fairness resource allocation algorithm in the long-term evolution wireless networks. Commun. IET **8**(17), 3105–3112 (2014)
11. Kim, C., Ford, R., Rangan, S.: Joint interference and user association optimization in cellular wireless networks. In: 2014 48th Asilomar Conference on Signals, Systems and Computers, Pacific Grove, CA, pp. 511–515 (2014)
12. Wu, J., Mehta, N., Molisch, A., Zhang, J.: Unified spectral efficiency analysis of cellular systems with channel-aware schedulers. IEEE Trans. Commun. **59**(12), pp. 3463–3474 (2011)
13. Gemici, O.F., Hokelek, I., Çırpan, H.A.: GA based multi-objective LTE scheduler. In: 2014 1st International Workshop on Cognitive Cellular Systems (CCS), pp. 1–5 (2014)

14. Kandukuri, S., Boyd, S.: Optimal power control in interference-limited fading wireless channels with outage-probability specifications. IEEE Trans. Wirel. Commun. **1**(1), 46–55 (2002)
15. Seo, H., Lee, B.G.: A Proportional Fair Power Allocation for Fair and Efficient Multiuser OFDM Systems, School of Electrical Engineering, Seoul National University, April 2004
16. Zhang, Y.J., Letaief, K.B.: Multiuser adaptive subcarrier-and-bit allocation with adaptive cell selection for OFDM systems. IEEE Trans. Wirel. Commun. **3**(4), 1566–1575 (2004)

Performance and Complexity Comparisons of Polar Codes and LDPC Codes

Mohammed Mensouri[✉] and Abdessadek Aaroud

Faculty of Sciences, Department of Computer Science, El Jadida, Morocco
mensourimohl@hotmail.com, a.aaroud@yahoo.fr

Abstract. Polar codes can be considered serious competitors to LDPC codes in terms of performance and complexity. This paper provides a description of the Polar codes and the LDPC codes used by channel coding. Then, we undertake a comparison of Polar codes and LDPC codes based on several factors: BER performance, encoding complexity and decoding computational complexity. The performance of newly obtained codes is evaluated in term of bit error rate (BER) for a given value of Eb/No. It has been shown via computer simulations. They are employed as the error correction scheme over Additive White Gaussian Channels (AWGN) by employing Binary phase shift keying (BPSK) modulation scheme.

Keywords: Channel coding · Polar codes · LDPC codes · Coding · Decoding · Successive cancellation algorithm · BP algorithm

1 Introduction

Polar Codes are linear block codes. Their invention is recent and proposed in [1]. These are the only codes for which one can mathematically prove that they reach the Shannon limit for an infinite size code. In addition, they have low coding and decoding complexity using a particular algorithm called Successive Cancellation (SC). The construction of the code as well as the coding and the decoding are explained in the different parts of this section before locating the performances of the Polar Code with respect to the existing codes.

Polar Codes can be used in a variety of practical communication scenarios. Indeed, they are faced with the Holevo capacity in [2] and the ITU G.975.1 standard in [3] for optical communications. In [4], the authors consider Polar Codes for multiple access channels to two users. The application of Polar Codes for voice communications is discussed in [5]. The authors compare performance with systems using LDPC codes for AWGN and Rayleigh channels. In [6], the authors consider a symmetric and memory-free relay channel, they show that the Polar Codes are suitable for compression-transmission applications. Polar Codes can be implemented for quantum communication channels from [7]. Security applications may use Polar Codes. The authors in [8] showed that Polar Codes asymptotically reach the capacity region of a wiretap channel under certain conditions.

In 1963, Gallager introduced a family of error correcting codes constructed from the matrix of low density parity, called LDPC (Low Density Parity Check Code) [9].

© Springer Nature Switzerland AG 2020
M. Ezziyyani (Ed.): AI2SD 2019, LNNS 92, pp. 207–216, 2020.
https://doi.org/10.1007/978-3-030-33103-0_21

These were forgotten with time, but these codes are good ideas particularly relevant operator for the construction of good codes. Thus, Gallager uses random permutations between parity codes to construct an efficient code of low complexity that imitates the random coding.

LDPC codes can be found in applications ranging from wireless Local/Metropolitan Area Networks LAN/MAN) (IEEE 802.11n [10] and 802.16e [11]) and high-speed wireless personal area networks (PAN) (IEEE 802.15.3c [12]) to Digital Video Broadcast (DVB-S2 [13]). Furthermore, these codes are currently being proposed for next generation cellular and mobile broadband systems.

This paper compares the error performance and complexity of Polar codes with message passing decoding and LDPC codes.

The paper is organized as follows. In Section 2, we provide the overview of Polar codes. In Section 3, we describe the encoding and the decoding method of LDPC codes. The main contribution of the paper is Section 4, where comparisons of Polar codes and LDPC codes based on several criteria are presented. Finally, we provide some conclusions in Section 5.

2   Polar Codes

2.1   Construction of Polar Codes

A Polar Code $PC(N, K)$ is a linear block code of size $N = 2^n$, with n is a natural integer, containing k bits of information. The generator matrix of the code is a sub-matrix of the nth power of Kronecker k, noted $F = k^{\otimes n}$:

$$k = \begin{pmatrix} 1 & 0 \\ 1 & 1 \end{pmatrix} \text{ and } F = \begin{pmatrix} k^{\otimes n-1} & 0_{n-1} \\ k^{\otimes n-1} & k^{\otimes n-1} \end{pmatrix} \tag{1}$$

The matrix F is composed of k lines. These k lines are chosen assuming a decoding Successive Cancellation (SC) which allows polarizing the error probability of the message bits. A Polar Code can be represented in matrix form from the matrix F and in the form of factor graph. The latter can be used for coding and decoding.

2.2   Process of Coding Polar Codes

Like any block code, the code words associated with Polar Codes are defined by a generative matrix G of dimension $(K \times N)$. This generating matrix G is obtained by removing $(N - K)$ rows of the matrix F. The equivalent encoding process then consists in multiplying a vector of size K by this matrix G. An alternative coding process consists of constructing a vector, denoted U, containing the K information bits and $N - K$ frozen bits set at 0. This vector is constructed in such a way that the information bits are located on the indices more reliable corresponding to K lines of $F^{\times n}$ previously selected. The corresponding code word X can then be calculated simply such that:

$$X = U \times F^{\times n} \tag{2}$$

A block code can be represented as a factor graph. In the case of the Polar Codes, we have seen that the construction of the generating matrix is recursive. It is then possible to show that the construction of the graph is also recursive. More generally, the factor graph of a Polar Code is presented in [14], of size N $= 2^n$ is composed of n stages of $\frac{N}{2}$ nodes of parity of degree 3 and $\frac{N}{2}$ nodes of variables of degree 3. The degree of a node represents its number of connections with other nodes. The factor graph can be used for coding and decoding.

2.3 Decoding of the Polar Codes by Successive Cancellation

The successive cancellation decoding algorithm can decode the Polar Codes. For the Polar Codes, a first algorithm has been proposed in [14] and is detailed in the following paragraph.

Once the message is transmitted through the communication channel, the noisy version, $Y = Y_0^{N-1} = [Y_0, Y_1, \ldots, Y_{N-1}]$, of the code word $X = X_0^{N-1} = [X_0, X_1, \ldots, X_{N-1}]$ is received. Arıkan showed in [1], as Polar codes reached the channel capacity under the assumption of a successive cancellation decoding. This decoding consists in estimating a bit u_i from the observation of the channel and the knowledge of the bits previously estimated. The value of the estimated bit is noted $\hat{u}_1$. Each sample Y_i is converted into a format called Likelihood Ratio (LR). These LRs, denoted $L_{i,n}$. During the decoding the different values $L_{i,j}(0 \leq j \leq n)$ are updated as well as the values $S_{i,j}$. The latter, called partial sums represent the recoding of the bits $\hat{u}_i$, as and when they are estimated. The particular sequencing of operations is explained below. First, it should be noted that the update of $L_{i,j}$ and partial sums $S_{i,j}$, can be calculated efficiently using the graphical representation of the Polar Codes.

To successively estimate each bit u_i, the decoder is based on the observation of the vector from the channel $\left(L \begin{pmatrix} N-1 \\ 0 \end{pmatrix}, n \right) = \left[L_{(0,n)}, \ldots, L_{(N-1,n)} \right]$ and previously estimated bits $\hat{u}_0^{i-1} = [\hat{u}_0, \ldots, \hat{u}_{i-1}]$. For this purpose, the decoder must calculate the values of LRs following:

$$L_{i,0} = \frac{Pr(Y_0^{N-1}, \hat{u}_0^{i-1} | u_i = 0)}{Pr(Y_0^{N-1}, \hat{u}_0^{i-1} | u_i = 1)} \tag{3}$$

During the decoding, when updating a stage $j > 0$, it is not the bits u_i that are used directly, but the partial sums, $S_{i,j}$ which are a combination of these estimated bits. When the decoding updates an LR of the stage $0, L_{i,0}$, then the decoder makes a decision as to the value of the bit ui such that:

$$\hat{u}_i = \begin{cases} 0 \ si \ L_{i,0} > 1 \\ 1 \ si \ non \end{cases} \tag{4}$$

The decoder knows the frozen bits. Therefore, if u_i is a frozen bit then $\hat{u}_i = 0$ regardless of the value of $L_{i,0}$. The decoder successively estimates the bits $\hat{u}_i$ from LRs, $L_{i,j}$ and partial sums $S_{i,j}$ which are calculated such as:

$$L_{i,j} = \begin{cases} F\left(L_{i,j+1}, L_{i+2^j,j+1}\right) & si\ B_{ij} = 0 \\ G\left(L_{i-2^j,j+1}, L_{i,j+1}, S_{i-2^j,j}\right) & si\ B_{ij} = 1 \end{cases} \tag{5}$$

$$S_{i,j} = \begin{cases} H_l\left(S_{i,j-1}, S_{i+2^{j-1},j-1}\right) & si\ B_{ij} = 0 \\ H_u\left(S_{i,j-1}\right) & si\ B_{ij} = 1 \end{cases} \tag{6}$$

With:

$$\begin{cases} F(a,b) = \frac{1+ab}{a+b} \\ G(a,b,S) = b \times a^{1-2S} \\ H_u(S) = S \\ H_l(S,St) = S \oplus St \\ B_{i,j} = \frac{i}{2^j} mod\ 2,\ 0 \le i < N\ and\ 0 \le j < n \end{cases} \tag{7}$$

The partial sum $S_{i,j}$ corresponds to the propagation of hard decisions in the factor graph. The decoding algorithm can alternatively be represented by a complete binary tree. The branches symbolize the functions F, G, H_u and H_l. Nodes represent LRs and intermediate partial sums, computed during the decoding process. The functions F and G process the LRs from right to left and store the results in the nodes on the left. The H functions retrieve partial sums from the left node, process them and store them in the right node.

3 LDPC Codes

3.1 Construction of LDPC Codes

LDPC code of parameters $(N, j, 1)$ is a linear block code of length N such that the parity check matrix H has j '1' in column and l '1' in a row. The numbers j and l are very small compared to the length of code to provide a low density matrix.

Figure 1 shows the parity check matrix of an LDPC code. This matrix H has therefore Nj/l rows, i, e. code is constituted Nj/l parity equation. The rate of the code then verifies $R \ge 1 - j/l$.

The parity check matrix H, as presented by Gallager in his thesis [9], can be divided into j sub-matrix $H^1, \ldots, H^j$, each containing an only '1' by column. The first sub matrix H^1 is a kind of identity matrix in which each '1' would be replaced by l '1', and whose number of columns is therefore multiplied by a factor l. The $j - 1$ other sub matrix $H^2, \ldots, H^j$ are obtained by applying $j - 1$ random permutation $\pi 2, \ldots, \pi j$ on the columns of the sub-matrix H^1. Thus the matrix of LDPC code with $(N = 20, j = 3, l = 4)$ is given in Fig. 2.

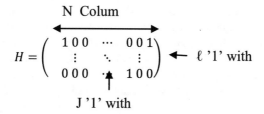

Fig. 1. Parity check matrix of an LDPC code (N, j, l)

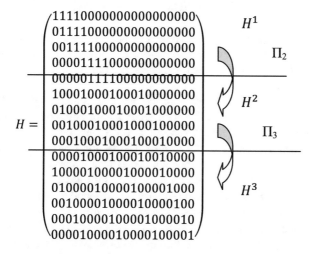

Fig. 2. Parity check matrix of an LDPC code (20, 3, 4)

3.2 Encoding Processes for LDPC Codes

The introduction of systematic LDPC codes by Tanner [14] was motivated primarily by reducing the encoding complexity. The problem of encoding LDPC codes is determined codeword x from the vector b containing the information bits and the parity check matrix H defined by the following equation:

$$Hx = 0 \tag{8}$$

If G is a generator matrix of the LDPC code to encode we have the following relationship:

$$x = Gb \tag{9}$$

Note that this equation introduced in (10) we obtain the following relationship:

$$HG = 0 \tag{10}$$

The method that seems to be simpler for encoding is to build a matrix encoding H′ upper triangular by applying the Gaussian elimination on the matrix H. As shown in Fig. 5. The x codeword is then cut into a systematic part containing the information bits, b_n with n = 1, ..., k and the part containing the redundant bits r_m with m = 1, ..., M determining. Then by identifying form the bottom up, it is easy to calculate the redundancy bits. However, the use of the Gauss pivot of the sparse matrix H only allows to obtain a semi-encoding matrix H′ upper triangular, with important density '1' making the number of the operation to be performed when the encoding of the order of N^2 while their direct competitor, that is to say, the turbo codes have a linear complexity $\theta(N)$. This encoding step is the principal drawback in the practical implementation of LDPC codes.

To reduce this complexity two approaches have been developed. The first is to constrain the matrix parity checking to reduce this complexity, while the second is based on the construction of a matrix encoding semi-triangular the low density as possible.

Regarding the first approach, we note the work of Spielman et al. [15, 16] which consist of concatenated LDPC codes with short length (sub-code) then the resulting code is incredible with linear complexity. However, given the size of each block (sub-code) is very small compared to the size of the resulting block, the performance of these types of codes are inferior to those obtained with a standard LDPC code of the same length. Another approach that was introduced by Luby et al. [17] is to force the parity check matrix to be symmetrical (either directly encoding) i.e. of the set of codes is limited by the fact that the matrix H either the semi-triangular as described in Fig. 3. Given that the parity matrix is low density, this restriction ensures linear complexity. It generates, in general, some small losses in performance. Methods of reducing the complexity of encoding in the form of the parity check matrix have been also developed in [18, 19].

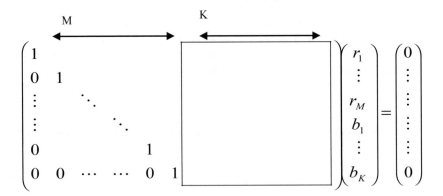

Fig. 3. Encoding LDPC codes

3.3   Decoding Processes for LDPC Codes

LDPC codes can be decoded iteratively by the sum of product algorithm, called the message passing algorithm the LDPC codes. The MAP decoding algorithm achieves performance, provided that the code graph is cycle free. Although it is closer to the algorithms decision decoding based on the work of Gallager [9]. The sum-product algorithm [20] is applied to the code graph where the check nodes and variable nodes exchange messages which mainly extrinsic reliability values associated with each code symbol. One can find other algorithms for decoding LDPC codes [21].

4   Comparisons of Polar Codes and Turbo Codes

In this section, we compare several aspects of coding and decoding Polar codes and LDPC codes. We plot the bit error probability versus the signal-to-noise ratio Eb/N0 to compare the BER-FER performance of Polar codes and LDPC codes by using the simulation resultants.

4.1   BER-FER Performance Comparison of Polar Codes and LDPC Codes

A comparison was made between an LDPC code and a Polar Code. In pure performance, Polar Codes are below LDPC codes for original SC decoding. A loss of about 0.9 dB was observed for a $BER = 10^{-3}$ in Fig. 4 between an LDPC code of size $N = 1056$, of yield $R = 1/2$ and performing $Imax = 50$ iterations, and a code Polar CP (1024, 512).

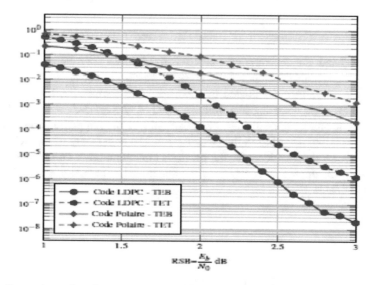

Fig. 4. Comparison of performances on a AWGN channel of a LDPC code and a Polar Code

The performance of a systematic Polar Code with a size 16 times larger, CP (16384, 8192), is then compared to that of the previous LDPC code in Fig. 5. It appears that in this configuration of equivalent computational complexity, the performance of the Polar Code is about 0, 2 dB higher for a $TEB = 10^{-4}$ compared to the LDPC code. This family of error correction code is therefore an alternative to the LDPC codes.

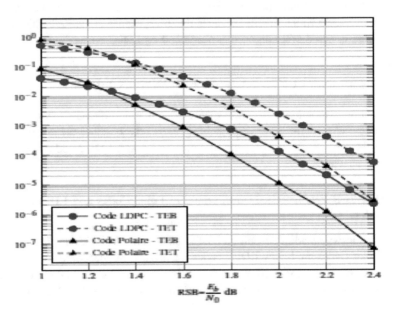

Fig. 5. Comparison of performances on a AWGN channel of a Turbo code and a Polar Code

4.2 Complexity Comparison of Polar Codes and LDPC Codes

On the other hand, the decoding complexity is much lower for a Polar Code, using an SC algorithm, systematic or not, for the same reasons as above. LDPC codes, like Turbo codes, are largely dominated by their decoding complexity. Indeed, the complexity of a BP type decoding algorithm is equal to $O(Imax(Ndv + Mdc))$. The variables dv and dc represent the average degree of the variable and parity nodes. These are equal to $dv = 4$ and $dc = 6:5$ in the present case. The complexity of the SC decoding algorithm is $O(Nn)$. Therefore, for the previous configuration, a Polar Code having a length 16 to 32 times greater provides computational complexity equivalent to the LDPC code (Table 1).

Table 1. comparison of computational complexity of coding and decoding of Polar Codes and LDPC codes

	Coding		Decoding	
Code	Structure	Complexity	Algorithm	Complexity
Polar code	Recursive encoder	$\sigma(Nn)$	Successive cancellation	$\sigma(Nn)$
LDPC code	Matrix multiplication	$\sigma(N^2)$	BP	$O(Imax(Ndv + Mdc)$

5   Conclusion

We have seen in this paper the construction, the coding and the decoding of the Polar Codes. The decoding algorithm used for Polar code is Successive Cancellation, which has been detailed because a lot of work derives from it. Then, we have presented an overview of LDPC code. Finally, a comparison of the decoding performances with other current codes, for an equivalent computational complexity, shows that the Polar Codes have performances that make it possible to compete with the LDPC codes.

References

1. Arikan, E.: Channel polarization: a method for constructing capacity-achieving codes. In: IEEE International Symposium on Information Theory, ISIT 2008, pp. 1173–1177 (2008)
2. Guha, S., Wilde, M.: Polar coding to achieve the Holevo capacity of a pure-loss optical channel. In: 2012 IEEE International Symposium on Information Theory Proceedings (ISIT), pp. 546–550 (2012)
3. Wu, Z., Lankl, B.: Polar codes for low-complexity forward error correction in optical access networks. In: ITG Symposium; Proceedings of Photonic Networks, vol. 15, pp. 1–8 (2014)
4. Sasoglu, E., Telatar, E., Yeh, E.: Polar codes for the two-user binary-input multiple-access channel. In: 2010 IEEE Information Theory Workshop (ITW), pp. 1–5 (2010)
5. Zhang, C., Yuan, B., Parhi, K.: Low-latency SC decoder architectures for polar codes arXiv: 1111.0705 (2011)
6. Lasco-serrano, R., Thobaben, R., Andersson, M., Rathi, V., Skoglund, M.: Polar codes for cooperative relaying. IEEE Trans. Commun. **60**, 3263–3273 (2012)
7. Hirche, C.: Polar codes in quantum information theory (2015)
8. Andersson, M., Rathi, V., Thobaben, R., Kliewer, J., Skoglund, M.: Nested polar codes for wiretap and relay channels. Commun. Lett. **14**, 752–754 (2010)
9. Gallager, R.G.: Low Density Parity Check Code. MIT Press, Combridge (1963)
10. IEEE-802.11n: Wireless LAN Medium Access Control and Physical Layer Specifications: Enhancements for Higher Throughput, P802.11n-2009, October 2009
11. IEEE-802.16e: Air Interface for Fixed and Mobile Broadband Wireless Access Systems, P802.16e-2005, October 2005
12. IEEE-802.15.3c: Amendment 2: Millimeter-wave-based Alternative Physical Layer Extension, 802.15.3c-2009 (2009)
13. ETSI DVB-S2: Digital video broadcasting, second generation, ETSIEN 302307, vol. 1.1.1 (2005)

14. Tanner, R.M.: A recursive approach to low complexity codes. IEEE Trans. Inform. Theory **27**(5), 533–547 (1981)
15. Mackay, D., Wilson, S.T., Davey, M.C.: Compression of constructions irregular codes. IEEE Trans. Commun. **47**(10), 1449–1454 (1999)
16. Spielman, D.: Linear-time encodable and decodable error-correcting code. IEEE Trans. Inf. Theory **42**(6), 1723–1731 (1969)
17. Luby, M., Mitzenmacher, M., Shokrollahi, A., Spielman, D., Stemann, V.: Partical loss Resilient codes. In: Proceedings 30th Annual ACM Symposium Theory of Computing, pp. 150–159 (1997)
18. Richardson, T.J., Urbanke, R.L.: Efficient encoding of low-density Parity check Code. IEEE Trans. Inf. Theory **47**, 638–656 (2001)
19. Debaynast, A., Declercq, D.: Gallager codes for mulliple access. In: Proceedings of the ISIT 02 Symposium, Lausanne, Suitzerland, July 2002
20. Amador, E.: Aspects des Décodeurs LDPC Optimisés pour la Basse Consommation. Thèse, Université TELECOM ParisTech, Soutenue le, 31 Mars 2011
21. Matteo GORGOGLIONE: Analyse et construction de codes LDPC non-binaires pour des canaux à évanouissement", thése, Université de Cergy Pontoise, Soutenue le, 25 October 2012

Planar Hexagonal Antenna Array for WLAN Applications

Taoufik Benyetho[1]([⊠]), Hamid Bennis[2], Jamal Zbitou[1], and Larbi El Abdellaoui[1]

[1] LMEET, FST of Settat, Hassan 1st University, Settat, Morocco
t.benyetho@gmail.com
[2] TIM Research Team, EST of Meknes, Moulay Ismail University, Meknes, Morocco

Abstract. In this work, a new antenna array developed for ISM 2.4 GHz applications is introduced. The structure is composed by associating four elementary antennas of hexagonal form using fractal geometry. The antennas are powered using the quarter-wave transformer. The designed structure has been optimized with CST Microwave Studio, it has a good input impedance matching in the validated band with a high gain of 8.4 dBi and an aperture angle around 60°. The total dimensions of the structure are 150 × 75 mm^2.

Keywords: Planar · Fractal · DGS · WLAN · Hexagonal · Antenna array

1 Introduction

Since 1886, when Heinrich Hertz proved by experiment that electromagnetic waves exist [1] and confirmed James Clerk Maxwell theory assuming that the magnetic and electric fields propagate in the air as waves, antennas have not ceased evolving leading to an evolution of wireless applications in this new century. After the first wireless transatlantic transmission between England and Canada in 1901 performed by Marconi in which he used a transmit antenna composed with vertical wires attached to the ground and a 200 m wire held up by a kite as a receiver antenna [2], the horn antenna [3] and the antenna arrays concept [4] were appeared. In 1972, the antenna world knew a breakthrough. Benefiting from the low-loss substrate production, micro strip antennas were developed. These structures present many advantages like the low cost of fabrication, the simplicity of production, the easiness of use within array structures and their simple incorporation with integrated circuits [5]. However they have some drawbacks like narrow bandwidth, low gain and large size. It exist some solutions to these disadvantages. The low gain problem could be resolved by using antenna array concept while the narrow bandwidth could be fixed by a DGS (defected ground structure) [6].

On other hand, applying fractal geometry could shrink the antenna size. Fractal objects are self-similar shapes that repeat themselves at different scales. The French mathematician Benoit Mandelbrot was the first who used the fractal term in his book "The fractal objects" [7] in 1975 even if the fractal geometry was known before Sierpinski triangle [8].

© Springer Nature Switzerland AG 2020
M. Ezziyyani (Ed.): AI2SD 2019, LNNS 92, pp. 217–225, 2020.
https://doi.org/10.1007/978-3-030-33103-0_22

Koch curve [9] or Minkowski curve [10] are all fractal forms that have emerged in the first half of the 20th century, but without any fields of application. It is thanks to Mandelbrot that fractal geometry has attracted the interest of scientists and since then many fields use this geometry. The first fractal antenna was published by Nathan Cohen in 1995 [11] and many progresses was done since then [12–15]. In this paper a new hexagonal planar antenna array based on fractal geometry is presented. Firstly, one element of the antenna array is fully studied. Then in the second part, the characteristics of the antenna array composed by four antenna elements are presented. The four antenna elements are fed using quarter wave transformer method. Finally, the results of the realized structure are shown and a comparison between this structure and others described where this work is standing.

2   One Element Antenna Design

The Fig. 1 shows the iteration 1 (a) and 2 (b) of the designed antenna. Iteration 2 consists in creating a hole having the same shape of the antenna at iteration 1 but with a different scale. The Table 1 summarizes the antennas dimensions.

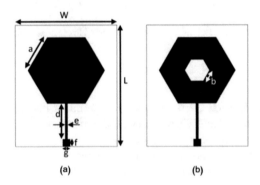

Fig. 1. First (a) and second (b) iterations of the designed antenna

In order to study the effect of the hexagonal hole over the structure, the antenna is simulated for different values of the parameter "b" relative to the size of the hexagonal hole. The Fig. 2 shows the reflection coefficients obtained.

If the diameter of the hexagonal hole widens, the resonance frequency of the antenna decreases against a poor impedance matching. It's deduced that the application of fractal geometry over this antenna gives it a shrinking ability. Indeed, 2.45 GHz of Fig. 2 is reached with a = 16 mm and b = 5 mm. For b = 7 mm, the resonance frequency is 2.3 GHz, in this case it is possible to shift the resonance frequency to 2.45 GHz by decreasing the value of the parameter "a" relative to the length of the side of the hexagon. But there are two problems to solve.

The first is the impedance mismatch in the resonance frequency, the second is the narrow bandwidth. The antenna is considered validated at 2.45 GHz only if the reflection

Table 1. Antennas dimensions (in mm).

Parameter	Length
L	50
W	42
a	16
b	5
d	14.6
e	1
f	3
g	3

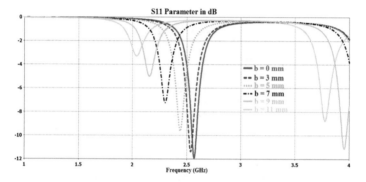

Fig. 2. Reflection coefficients for different "b" values

coefficient is less than −10 dB in all the band (2.4–2.5) GHz. One solution to these disadvantages is the use of a defected ground (DGS). This solution has shown good results for enlarging bandwidth and improving impedance matching [16–18].

Figure 3 illustrates the modified ground plane of the antenna by creating a hexagonal hole while Fig. 4 shows the antenna reflection coefficients for different values of the parameter "c" relative to the length of the hexagonal hole side of the ground plane. The other parameters of the antenna are those listed in Table 1.

It's remarked that the defected ground improves the bandwidth of the antenna and its impedance matching. When the dimension of the hexagonal hole of the ground decreases, the resonant frequency also decreases to the detriment of the impedance matching. When it becomes much larger than the size of the antenna (c = 20 mm), the impedance matching drops rapidly.

What is deduced is that when creating a hole in the radiating patch, the resonant frequency decreases by increasing the size of the hole but the impedance matching is degraded. When creating a hole in the ground plane, the resonance frequency decreases by decreasing the hole size but the impedance matching is degraded. In order to define the best parameters of the antenna, the CST Microwave Studio optimization tool was used.

The goal is to achieve a reflection coefficient of −30 dB at 2.45 GHz. The Fig. 5 shows the reflection coefficient obtained after the optimization while the Table 2 shows the final dimensions of the antenna in mm.

Fig. 3. The designed antenna ground

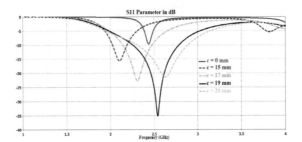

Fig. 4. Reflection coefficients for different "c" values

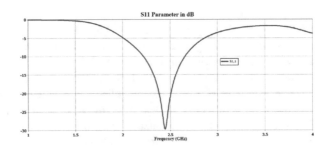

Fig. 5. The optimized reflection coefficient of one element antenna

Table 2. Optimized antenna dimensions (in mm).

Parameter	Length
a	16
b	5
c	18

The antenna bandwidth at −10 dB is 2.22 GHz to 2.65 GHz with a value of −29.65 dB at the resonance frequency (2.45 GHz). This bandwidth (17.5%) fully covers the 2.4 GHz ISM band defined from 2.4 to 2.5 GHz. Figure 6 shows the radiation pattern of the antenna in the XZ and YZ planes at 2.45 GHz (left) and its current distribution (right). The antenna presents an omnidirectional radiation in the XZ plane. In the YZ plane it is bidirectional along the Z axis (0° and 180°) with a gain of 3.4 dBi and an aperture angle of 78°. The majority of the current is distributed on the left and right edges of the hexagon.

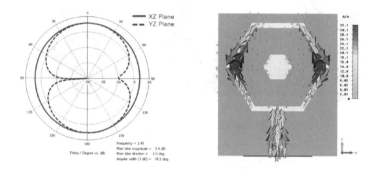

Fig. 6. The antenna radiation pattern (left) and current distribution (right)

3 Antenna Array Design

After the validation of the elementary antenna, an antenna array composed by four (4) elementary antennas to increase the gain of the structure is designed. Figure 7 shows the top (a) and the back (b) faces of the antenna array while Table 3 summarizes its dimensions.

After many optimizations, the parameters allowing to have the best reflection coefficient at the 2.45 GHz resonance frequency are obtained. In order to validate this optimization the structure were simulated in both ADS and CST electromagnetic simulators. Figure 8 shows the reflection coefficients obtained by the two software (left).

The bandwidth at −10 dB obtained from both software is almost the same (2.22–2.65) GHz, which covers the 2.4 GHz ISM band. The difference between the two software lies in the value of the reflection coefficient at the resonant frequency. While ADS has a very good value of −33 dB, the best result was −18 dB for CST. This difference is due to the algorithms used by each software.

The current distribution of the antenna array remains identical to that of an elementary antenna. On the other hand, the radiation pattern has naturally become more directive. Figure 8 depicts the radiation pattern of the antenna array at the 2.45 GHz in XZ and YZ planes (right).

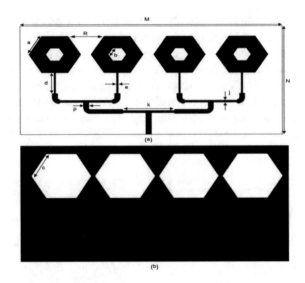

Fig. 7. The antenna array top (a) and back (b) faces

Table 3. Antenna array dimensions (in mm).

Parameter	Length
M	150
N	75
j	1.55
k	29.2
p	3
R	20.5

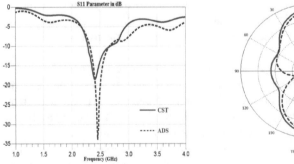
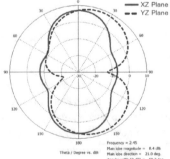

Fig. 8. The Reflection coefficients (left) and radiation pattern (right) of the designed antenna array

In the XZ plane, the antenna has passed from omnidirectional radiation to bidirectional radiation along the Z axis (0° and 180°).

In the YZ plane, the antenna has kept its bidirectional character but it is shifted 20° to the right of the Z axis (20° and 160°). The gain increased to 8.4 dBi and the aperture angle became 60°.

4   Results and Discussion

Figure 9 shows a picture of the realized structure (left) and a comparison of the reflection coefficient (middle) and the radiation pattern (right) at 2.45 GHz to the simulated antenna array results.

If CST displays a bandwidth (2.22–2.65) GHz with 2.4 GHz resonance frequency, the measurements show a tri-band network, with resonant frequencies at 2.28 GHz, 2.66 GHz and 3.47 GHz.

In terms of bandwidth set at −10 dB, the measurements validate two bands (2.2–2.8) GHz and (3.4–3.55) GHz. Although the measurement validated bands completely cover the 2.4 GHz ISM band, the difference in results between the simulation and the measurement required an analysis which showed a slight deformation in one of the lines feeding an antenna array element during the realization process.

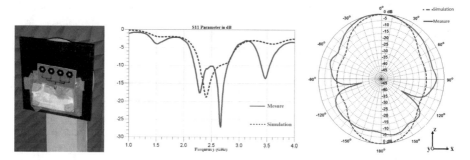

Fig. 9. Photo of the realized antenna array (left), its reflection coefficient (middle) and its radiation pattern (right)

Like the simulated radiation pattern, the measured one also has two directions along the Z axis at 2.45 GHz. Table 4 presents a comparison of the characteristics of the validated antenna with other structures.

Table 4. Comparison of validated antenna array with other structures.

Structure	Dimensions (mm^2)	Bands (GHz)	Gain (dBi)
Proposed antenna array	150 × 75	2.5	8.4
		3.5	8.4
Reference [19]	226 × 136	2.4	14.5
		3.5	16.5
Reference [20]	306 × 306	2.45	13.2
Reference [21]	200 × 50	2.4	4.6
		3.5	7
		5.2	7.9

5   Conclusion

A new hexagonal antenna array based on fractal geometry and covering the 2.4 GHz ISM bands is developed. The structure presents good impedance matching, gain and aperture angle in the validated band. Despite the difference between the results of the simulated and measured reflection coefficients of the antenna array, the main characteristics remained satisfactory with a bandwidth covering the target ISM band and an important gain compared to the structure dimensions, which makes it suitable for wireless application use like WLAN and RECTENNA. Also it's possible to match the same structure to other frequencies without altering its characteristics by changing the antenna elements dimensions. Some future works could be applying more fractal iterations to the antenna elements and see how that's affect the structure results. It's also possible to use more than four antenna elements in order to increase the gain.

Acknowledgments. The authors have to thank Mr. Mohamed Latrach, Professor in ESEO, Engineering Institute in Angers, France, for allowing them to use all the equipment and electromagnetic solvers available in his laboratory.

References

1. Van Name, F.W.: Modern Physics, p. 30. Prentice-Hall, New York (1962)
2. Belrose, J.S.: Fessenden and Marconi: their differing technologies and transatlantic experiments during the first decade of this century. In: International Conference on 100 Years of Radio, pp. 5–7 (1995)
3. Olver, D.: Microwave horns and feeds, USA, IET, pp. 2–4 (1994). ISBN 0-85296-809-4
4. Spradley, J.: A volumetric electrically scanned two-dimensional microwave antenna array. IRE National Convention Record, Part I - Antennas and Propagation Microwaves. The Institute of Radio Engineers, New York, pp. 204–212 (1958)
5. Howell, J.: Microstrip Antennas. In: IEEE International Symposium on Antennas and Propagation. Williamsburg Virginia, pp. 177–180 (1972)
6. Bancroft, R.: Microstrip and Printed Antenna Design. Noble Publishing, Chapter 2–3 (2004)
7. Mandelbrot, B.B.: Fractals: form, chance and dimension. Les objets fractals: forme hasard et dimension. Nouvelle bibliothèque scientifiques. French edition (1975)

8. Puente, C., Romeu, J., Cardama, R.: On the behavior of the Sierpinski multiband fractal antenna. IEEE Trans. Antenna Propag. **46**(4) (1998)
9. Koch, H.V.: Une méthode géométrique élémentaire pour l'étude de certaines questions de la théorie des courbes planes. Acta Mathematica **30**(1), 145–174 (1906)
10. Dhar, S., Patra, K., Ghatak, R., Gupta, B., Poddar, D.R.: A dielectric resonator-loaded minkowski fractal-shaped slot loop heptaband antenna. IEEE Trans. Antennas Propag. **63**(4), 1521–1529 (2015)
11. Cohen, N.: Fractal Antennas. Commun. Q. **9** (1995)
12. Jindal, S., Sivia, J.S., Bindra, H.S.: Hybrid fractal antenna using meander and minkowski curves for wireless applications. Wireless Pers. Commun. https://doi.org/10.1007/s11277-019-06622-5
13. Sivia, J.S., Kaur, G., Sarao, A.K.: A modified Sierpinski carpet fractal antenna for multiband application. Wireless Pers. Commun. **95**(4), 4269–4279 (2017)
14. Bhatia, S.S., Sivia, J.S.: On the design of fractal antenna array for multiband applications. J. Inst. Eng. India Series B (2019). https://doi.org/10.1007/s40031-019-00409-9
15. Benyetho, T., Zbitou, J., El Abdellaoui, L., Bennis, H., Tribak, A.: A new dual band planar fractal antenna for UMTS and ISM bands. Act. Passive Electron. Compon. (Hindawi) **2018**, 1–10 (2018)
16. Lu, P., Yang, X.S., Li, J.L., Wang, B.Z.: A compact frequency reconfigurable Rectenna for 5.2 and 5.8 GHz wireless power transmission. IEEE Trans. Power Electron. **30**(11), 1–5 (2015)
17. Benyetho, T., El Abdellaoui, L., Tajmouati, A., Tribak, A., Latrach, M.: Design of new Microstrip multiband fractal antennas: Sierpinski triangle and hexagonal structures. In: Handbook of Research on Advanced Trends in Microwave and Communication Engineering, pp. 1–33. IGI Global (2016). https://doi.org/10.4018/978-1-5225-0773-4.ch001
18. Benyetho, T., El Abdellaoui, L., Zbitou, J., Bennis, H., Tribak, A., Latrach, M.: A new dual band planar fractal antenna for UMTS and ISM bands. Int. J. Commun. Antenna Propag. (I. Re.C.A.P) **7**(1), 64–71 (2017)
19. Wang, Z., Zhang, G.X., Yin, Y., Wu, J.: Design of a dual-band high-gain antenna array for WLAN and WiMAX base station. IEEE Antennas Wirel. Propag. Lett. **13**, 1721–1724 (2014)
20. Fung, C.: Basic Antenna Theory and Application. Worcester Polytechnic Institute (2011)
21. Sonkki, M., Pfeil, D., Hovinen, V., Dandekar, K.R.: Wideband planar four element linear antenna array. IEEE Antennas Wirel. Propag. Lett. **13**, 1663–1666 (2014)

Services Search Techniques Architecture for the Internet of Things

Soukaina Bouarourou[1(✉)], Abdelhak Boulaalam[2],
and El Habib Nfaoui[1]

[1] LIIAN Laboratory, Sidi Mohamed Ben Abdellah University, Fez, Morocco
bouarourou.soukaina@gmail.com,
soukaina.bouarourou@usmba.ac.ma
[2] ENSA, Sidi Mohamed Ben Abdellah University, Fez, Morocco

Abstract. As we are moving towards the Internet of Things (IoT) with web services and cloud computing, we will have thousands of connected sensors and their data to handle and benefit from their services. With the enormous number of sensors available in the IoT environment, effectively and efficiently searching and selecting the best sensors regarding the user's requirement has recently become a crucial challenge. In this paper, we propose an effective context-aware method to cluster sensors in three categories in the form of the Semantic category (SC). Firstly, sensors of each SC are grouped based on their type to create Semantic Type Sensor Network (SSTN), in which sensors with similar context information are gathered into one cluster.

Keywords: IoT · Sensors · Searching and selecting · Context-aware · Context information · User's requirement

1 Introduction

The physical world is being increasingly penetrated by embedded sensors that are connected to the internet with the goal of enhancing human life. This is the driving force behind the internet of things (IoT), A thing within the Internet of Things; it can be someone with a cardiac monitor implant, an automobile that has inbuilt sensors which will alert the driver when tire pressure is low or any other natural or man-made object. The main aim of IoT is to create a better world for people, where things around us communicate with one another and know what we like, what we want, and what we need and act accordingly without specific instructions [1–3], Fig. 1 [4] illustrates the definition more clearly. with the IoT, the world web becomes more sensitive to the senses, for example: monitoring all sorts of environmental phenomena (weather, noise, light ....) [5].

IoT consists of three layers, such as the application layer, network layer, and sensing layer. The sensing layer consists of various sensing devices, such as controllers, sensor devices (smoke, temperature, pressure, light, and humidity), RFID tags, camera, energy meter, and actuators, etc. which have limited energy, memory and processing power, presented in Fig. 2 [6].

© Springer Nature Switzerland AG 2020
M. Ezziyyani (Ed.): AI2SD 2019, LNNS 92, pp. 226–236, 2020.
https://doi.org/10.1007/978-3-030-33103-0_23

As in traditional internet, a search is key service in the IoT that enabling users to find real-world entities that exhibit a certain current state (the number of available bicycles at a rental stations after 6 pm, another example when monitoring, detecting or discovering phenomenon which sensors should be selected to retrieve the desired data?).

Given the increasing number of the sensors available and huge amount of data generated by IoT from diverse objects, this raises the challenge of searching and selecting the right sensors for a query in an efficient and effective way, which creates a need for searching effective methods in order to search the sensors and the data provided by them. Most of the existing approaches take into account the distributed nature and dynamicity of IoT systems. The whole objective of this work is to provide an efficient way for users to select appropriate sensors or services (Needs) with minimal time consumption.

The rest of this paper is organized as follows: Sect. 2 gives a review of the related work on sensor search and selection with summaries tables of all research share it. The proposed approach and example of our system are presented in Sect. 3. Finally, we present a conclusion and suggestions for further research.

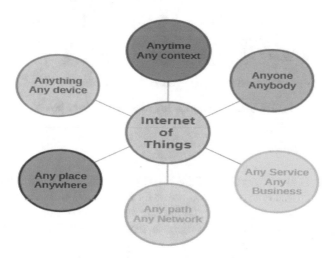

Fig. 1. Definition of the internet of things

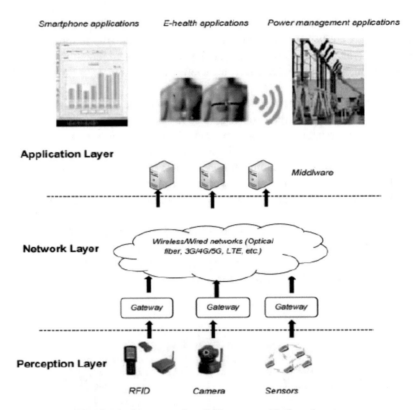

Fig. 2. Architecture of an IoT system with three layers.

2 Related Work

In different fields like Smart Cities, Smart Agriculture and environment monitoring in IoT, searching the sensors (service) is the main key operation. When users submit a query, IoT middleware solutions should provide the relevant sensor (service) data back to them quickly without asking the users to manually select the sensors which are relevant to their requirements.

By browsing literature, Methods that have been proposed to address the sensor search issues can be divided into:

- Research focused on processing data: content-based sensors.
- Research using context information sensors: context-aware.

Several approaches have been proposed for content-based sensor search by processing data generated by sensors and the clustering was based on it. The idea through in such work was to obtain the optimal subset of devices that provide desired data at any instance of time.

Bu et al. [7] propose the Fuzzy-c-means approach based on the tensor canonical polyadic decomposition to cluster IoT big data for drilling smart data, in particular, the tensor canonical polyadic decomposition is employed to reduce the attributes of each object in raw dataset significantly. Another such approach [8] used data-level fusion with PCA (Principal component analysis) to reduce the cost and keep the original information with the mini-batch partition-based clustering method combined with the algorithm DGO (dynamic Group Optimization).

Each big data has a large number of attributes, the chosen approach using to reduce it, not always be effective to keep original information.

In the other hand Elahi et al. [9] presented this method (content-based) sensor search which matches the user's query based on the estimated probability. The basic idea is to build prediction models by studying the relation between sensors. This models can predict which sensors would output the sought value at the time to the user; depend on the behavior of the people which is not periodic. This was one of the challenges that the approach faced.

Another such approach created a real-time search engine for the Internet named Dyser [10]. This engine uses statistical models to make predictions about the state of its registered sensors in the network and to search the entities in real-time, prediction models help to find selecting sensors with a minimum number of sensor data retrievals.

Truong et al. [11] also propose an algorithm, that exploits the fuzzy sets for calculating a resemblance score for any two sensors that were used to get a scored list of similar devices (sensors). This similarity was calculated using the sensor range difference, Although this approach was also content-based, it can be used in some applications, for instance; monitoring whether sensors are functioning properly, identifying sensors that need maintenance, deploying new sensors, or defining new services based on particular sensors.

Due to the previous impact of the content-based approach, other research proposes a context-based approach. Context-based will play a critical role in deciding what data needs to be proposed and much more [12]. Context is any information or properties, that can be used to characterize the situation of a sensor, and it is supposed that it would be provided by the sensors' manufacturers.

Some work exists have focused on sensor searches based on context properties. In [13], they have developed the Linked Stream Middleware (LSM) a platform that brings together the "live" data and web data. Despite this potential, this Middleware has limitations like; sensor selection based only on the type of the sensors (weather, color, smoke, Gas, traffic …), and pay a little attention to the sensor's attributes (availability, energy …) related to LSM, GSN [14], and Microsoft Sensor Map [15], are also created. GSN is a system for Internet-based interconnection of heterogeneous sensors, and sensor networks, supporting homogeneous data-stream query processing on the resulting global set of sensor data streams. In GSN, a list of sensors will be provided for the user to manually select the most appropriate one. In Microsoft Sensor Map, users will be allowed to select sensors by using a location map, sensor type and keywords.

A context-aware sensor search, selection, and ranking model, called CASSARAM [16], to address the challenge of efficiently selecting a subset of relevant sensors out of a large set of sensors with similar functionality and capabilities. CASSARAM considers user preferences and a broad range of sensor characteristics such as reliability, accuracy, location, battery life, and many more. Moreover, CASSARAM did not consider the change that occurs in IoT, in other words, they ignored the performance of their system in dealing with the change (deleting or joining new sensors) in the sensor network. An effective context-aware method to cluster sensors is proposed in the form of sensor semantic overlay networks (SSONs) [17], in which sensors with similar context information are gathered into one cluster, firstly sensors are grouped based on their type to create sensor Semantic Overlay Networks (SSON), for changing sensor networks; an adaptive strategy is proposed to maintain the performance against dynamicity in the IoT.

The iterations of the AntClust algorithm increase in a linear way as the sensor count increases. The accuracy of the search result is very less when compared with CASSARAM.

Table 1. Different sensor/service search approaches and experimental evaluations.

Approach	Research	nb_of_sensor	Attribute/Property	Execution time	Accuracy	Dynamicity (add or delete)
Content-based sensor	Bu [5]	Unlimited	Unlimited	650 min/nb-clusters	0.90/nb-clusters	✓
	Tang [6]	Unlimited (12000)	100	0.05 min	✓	✓
	Elahi [7]	250	_	_	_	✓
	Dyser [8]	385 (limited)	_	_	_	_
	Truong [9]	42 (limited)	_			×
Context-aware	LSM [11], GSN [12, 13]	100 000	Limited (energy, location)			
	CASSARAM [14]	1000000	Unlimited	✓	✓	×
	SSON(s) [15]	1000000	Unlimited	×	×	✓

Table 1 summarizes the different research efforts that have addressed the challenge of sensor search. Table 1 lists the efforts and the number of sensors, execution time, attributes or properties, accuracy, dynamicity (e.g. deleting or joining new sensors in the sensor network) used in their experiments.

The Context property is any information can be used to characterize and identify the sensors in IoT; it was some common properties between sensors other than suggested in [16, 17].

The work cited above uses the clustering heuristics method to filter and select the sensor in question. In addition, it takes more time of clustering and research, most of the existing work requires users to set a predicted number of sensors as a result.

Considering different parameters such as the number of sensors, the processing time and the number of selected sensors. Having been inspired by the main concept discussed in [16, 17], we propose a new architecture system based on the clustering of existing sensors in the network into three categories. In comparison to the relevant works mentioned above, our proposed approach has several advantages such as; the using of the context-awareness to improve the search efficiency, the adopting of three categories of IoT, which include Environment, industry, and society. the employment of these will improve the efficiency of selecting the best sensors following the user's request regarding the quality and time of the proposed solution.

3 The Proposed System Architecture

This paper aims to allow users to search and retrieve information from sensors according to their need. As the physical layer of IoT consists of numerous sensors as shown in Fig. 2, when the user submit the query, The user enters the query at the application layer which is then decomposed and prioritized based on the set of predefined rules. Then the prioritized sub query is transformed into a low-level form as shown in Fig. 3. (Decomposed query) and then sent to the base station as given in Algorithm 1.

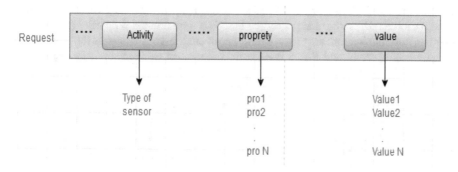

Fig. 3. Query decomposition.

Having been inspired by [4], the application domain can be mainly divided into three Semantic Categories (SC) based on their focus [18, 19]: industry, environment, and society. For example; Supply chain management [20], transportation and logistics [21], aerospace, aviation, and automotive are some of the industry-focused applications of IoT. Telecommunication, medical technology [22], healthcare, smart building, home [23] and office, media, entertainment, and ticketing are some of the society focused applications of IoT. Agriculture and breeding [24, 25], recycling, disaster alerting, environmental monitoring is some of the environment-focused applications.

The proposed Algorithm: Sensor Search Using Clustering Technique

```
Input: Query from user, Set of grouped   Sensors
          type - keywords and predefined set of
          rules.
Output: sensors (services) requested by user.
     1: Decompose the given query into words.
     2: Check for potential keywords
     3: Classify each keyword on:
        Activity, properties, property-keyword.
     4: specify the type of request
     5: Assign priorities to each property.
     6: Create a table in priority and Corresponding
        identifier
     7: Send it to base station- server.
     8: Search the available sensors based on the each
        Range which each property belong it
     9: Return value to the user
```

Next, we create in each SC, Semantic Type Sensor Network (STSN) in which sensors that have similar types (traffic, road activity, weather, etc.) are grouped together. In each STSN, sensors are clustered based on their context properties (availability, location, communication, accuracy, energy, cost, etc.), Figure 4 depicts the overview of our proposed system, and Fig. 5 depicts the component of semantic category.

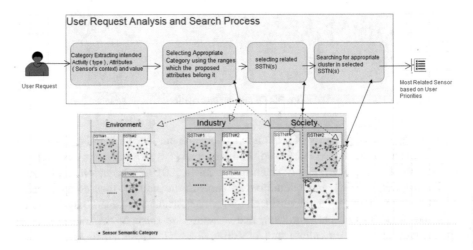

Fig. 4. Overview of the proposed system.

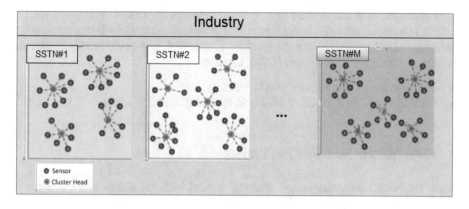

Fig. 5. The components of the semantic category.

Our model divides the user's query into two categories:

- Generalized query: it contains at least two properties.
- Specific query: it contains more than three properties.

When user's query is an as generalized query or Specific query, our system should be directed for intended SC, searching appropriate STSN Then, the combination of most relevant sensor clusters will be chosen from the selected STSNs to provide the best service to the user by respecting this priority of common properties between devices:

a. Availability: the existence of this sensor on real-time.
b. User's requirements: other properties that the user has marked in his query with a priority of each of them if it exists (Generalized query).

As an example, Fig. 6 shows a scenario for searching "TemperatureSensor".

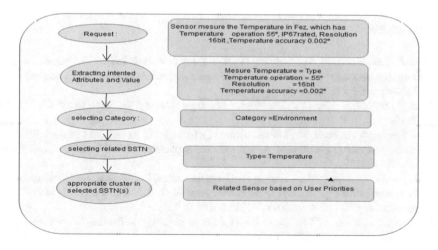

Fig. 6. Scenario searching TemperatureSensor

For the use case study, there is no existing operational 'Things' search system. For this, in this paper, the experimental part, it will be presented in the future work where the range of the properties of each category is specified. For example, most industrial sensors are built with specifications that include:

- Temperature operation from −40 to +85°.
- Resolution of 12 or 16 bit, enabling temperature accuracy down to .01° or less.

4   Conclusions and Future Trends

Since the sensor technologies are growing around IoT, the search and discovery of resources are one of the main challenges in IoT environments. In this work, we have analyzed, compared the most past research work and discuss their applicability towards the IoT. For that, we have presented a new architecture system to cluster all exist sensors in three SC (environment, society, industry) based on the context information of the sensor to search and select the most related sensor based on the user query.

The proposed work in progress has some limitations such as:

- List of specifications with a range of each category.
- Semantic model contains the most designative properties such as: location, communication (wireless personal area network (WPAN) (e.g. Bluetooth), wireless local area network (WLAN) (e.g. Wi-Fi), and wireless metropolitan area network (WMAN) (e.g. WiMAX), wireless wide area network (WWAN) (e.g. 2G and 3G networks), and satellite network (e.g. GPS)), user's requirement (cost, energy, availability, accuracy ....).
- Applicate efficient sensor clustering Algorithm.

We intend to explore several of the previous mentions optimizations and improve our model.

References

1. Fredj, S.B., Boussard, M., Kofman, D., Noirie, L.: A scalable IoT service search based on clustering and aggregation. In: IEEE International Conference on Green Computing and Communications (2013)
2. Li, S., Da Xu, L., Zhao, S.: The internet of things: a survey. Inf. Syst. Front. **17**, 243–259 (2015)
3. Al Fuqaha, A., Guizani, M., Mohammadi, M., Aledhari, M., Ayyash, M.: Internet of Things: a survey on enabling technologies, protocols and applications. IEEE Commun. Surv. Tutor., 601–642 (2015)
4. Perera, C., Zaslavsky, A., Christen, P., Georgakopoulos, D.: Context aware computing for the internet of things: a survey. IEEE Commun. Surv. Tutorials **16**(1), 414–454 (2014). https://doi.org/10.1109/surv.2013.042313.00197
5. Cooper, J., James, A.: Challenges for database management in the internet of things. IETE Tech. Rev. **26**(5), 320–329 (2009)

6. Abdmeziem, M.R., Tandjaoui, D., Romdhani, I.: Architecting the internet of things: state of the art. In: Robots and Sensor Clouds, pp. 55–75. Springer, Cham (2016)
7. Bu, F.: An efficient fuzzy c-means approach based on canonical polyadic decomposition for clustering big data in IoT, 675–682 (2018)
8. Tang, R., Fong, S.: Clustering big IoT data by meta heuristic optimized mini-batch and parallel partition-based DGC in Hadoop, March 2018
9. Elahi, B.M., Romer, K., Ostermaier, B., Fahrmair, M., Kellerer, W.: Sensor ranking: a primitive for efficient content-based sensor search. In: Proceedings of the 2009 International Conference on Information Processing in Sensor Networks, pp. 217–228. IEEE Computer Society (2009)
10. Ostermaier, B., Romer, K., Mattern, F., Fahrmair, M., Kellerer, W.: A real-time search engine for the web of things. In: Internet of Things (IOT), pp. 1–8. IEEE (2010)
11. Truong, C., Romer, K., Chen, K.: Fuzzy-based sensor search in the web of things. In: 2012 3rd International Conference on the Internet of Things (IOT), pp. 127–134. IEEE (2012)
12. Perera, C., Zaslavsky, A., Christen, P., Georgakopoulos, D.: Context aware computing for the internet of things: a survey. IEEE Commun. Surv. Tutor. **16**(1), 414–454 (2014)
13. Le-Phuoc, D., Quoc, H.N.M., Parreira, J.X., Hauswirth, M.: The linked sensor middleware-connecting the real world and the semantic web. In: Proceedings of the Semantic Web Challenge, vol. 152 (2012)
14. Aberer, K., Hauswirth, M., Salehi, A.: Infrastructure for data processing in large-scale interconnected sensor networks. In: 2007 International Conference on Mobile Data Management, pp. 198–205. IEEE (2007)
15. Nath, S., Liu, J., Zhao, F.: Sensor map for wide-area sensor webs. Computer **40**(7), 90–93 (2007)
16. Perera, C., Zaslavsky, A., Liu, C.H., Compton, M., Christen, P., Georgakopoulos, D.: Sensor search techniques for sensing as a service architecture for the internet of things. IEEE Sens. J. **14**(2), 406–420 (2014)
17. Ebrahimi, M., Shafieibavani, E., Wong, R.K., Chi, C.H.: A new meta-heuristic approach for efficient search in the Internet of Things. In: IEEE International Conference on Services Computing (2015)
18. Atzori, L., Iera, A., Morabito, G.: The internet of things: a survey. Comput. Netw. **54**(15), 2787–2805 (2010). http://dx.doi.org/10.1016/j.comnet.2010.05.010
19. Sundmaeker, H., Guillemin, P., Friess, P., Woelffle, S.: Vision and challenges for realizing the internet of things. European Commission Information Society and Media, Technical report, March 2010. http://www.internet-of-things-research.eu/pdf/IoTClusterbookMarch2010.pdf. Accessed 10 Oct 2011
20. Chaves, L.W.F., Decker, C.: A survey on organic smart labels for the internet-of-things. In: 2010 Seventh International Conference on Networked Sensing Systems (INSS), pp. 161–164 (2010). http://dx.doi.org/10.1109/INSS.2010.5573467
21. Chen, Y., Guo, J., Hu, X.: The research of internet of things' supporting technologies which face the logistics industry. In: 2010 International Conference on Computational Intelligence and Security (CIS), pp. 659–663 (2010). http://dx.doi.org/10.1109/CIS.2010.148
22. Wang, Y.-W., Yu, H.-L., Li, Y.: Internet of things technology applied in medical information. In: 2011 International Conference on Consumer Electronics, Communications and Networks (CECNet), pp. 430–433, April 2011. http://dx.doi.org/10.1109/CECNET.2011.5768647
23. Chong, G., Zhihao, L., Yifeng, Y.: The research and implement of smart home system based on internet of things. In: 2011 International Conference on Electronics, Communications and Control (ICECC), pp. 2944–2947, September 2011. http://dx.doi.org/10.1109/ICECC.2011.6066672

24. Burrell, J., Brooke, T., Beckwith, R.: Vineyard computing: sensor networks in agricultural production. IEEE Pervasive Comput. **3**(1), 38–45 (2004). http://dx.doi.org/10.1109/MPRV. 2004.1269130
25. Lin, L.: Application of the internet of thing in green agricultural products supply chain management. In: 2011 International Conference on Intelligent Computation Technology and Automation (ICICTA), vol. 1, pp. 1022–1025 (2011). http://dx.doi.org/10.1109/ICICTA. 2011.256

Smart Geoportal for Efficient Governance: A Case Study Municipality of M'diq

Khalid Echlouchi[1(✉)], Mustapha Ouardouz[1],
Abdes-Samed Bernoussi[2], and Hakim Boulaassal[2]

[1] Faculty of Sciences and Techniques,
Mathematic Modeling and Control (MMC), Tangier, Morocco
khalid.echlouchi.doc@gmail.com, ouardouz@gmail.com
[2] Faculty of Sciences and Techniques,
Geoinformation and Land Management (GAT), Tangier, Morocco

Abstract. The land management has experienced the implementation of innovative new digital tools to enhance urban development in cities and improve their governance. The Smart governance is based on the federation of new technologies while allowing interoperability between different urban services, and offering free access to data and information. It is based on a close and real-time interactions between citizens and decision-makers to ensure efficient urban services while allowing the capitalization of skills. To do that, digital tools like GIS platforms and Geoportals, have been widely used to improve the performance of public services and the degree of citizen satisfaction.

In this paper we consider a smart Geoportal project of the city of M'diq as a decision-making platform covering: Territorial data layers, 3D model and BI management tools for municipal decision-making.

This platform is a demonstration of the use of Geoportals for smart governance in the implementation of smart city projects that requires complex interactions between government, citizens and public services, which are simplified through those tools.

Keywords: Smart city · Smart governance · Geoportal · Web-GIS · Public services · Decision-making

1 Introduction

To control the planning and development of a territory and achieve high governance, local authorities and public services need precise data and cartography which can only be achieved through innovative digital tools allowing data gathering, sharing, observation, spatial analysis, modeling, displaying large amounts of data and decision-making [1].

Therefore, the use of new technologies promotes the transformation of the traditional city into a smart city, which requires the adoption and implementation of new concepts such as Smart Economy, Smart Mobility, Smart Environment, Smart People, Smart Living and Smart Governance [2].

© Springer Nature Switzerland AG 2020
M. Ezziyyani (Ed.): AI2SD 2019, LNNS 92, pp. 237–245, 2020.
https://doi.org/10.1007/978-3-030-33103-0_24

Smart Governance, one of the main pillars of the smart city, is based on the federation of new technologies while allowing interoperability between the different public services, besides of the free access to data and information. One of its principal pillar is the interaction and exchange between citizens and decision-makers to ensure the performance of public services while allowing the capitalization of skills [2, 3].

To concretize this local Smart Governance and carry out a complete cartography of the intervention territory, local authorities and public services are invited to set up a shared spatial data infrastructure with the main objective to collect and harmonize the exchange of geographic data at the level of a unit of territory (municipality, province, region …) between services and to encourage exchanges with the local population. The availability of such data: topographic, city planning, infrastructure (water, sanitation, electricity, roads), cadastral data, environmental data, risk area, … will help stakeholders improve their decisions [4].

In this article, we present our experience of setting up a Geoportal "M'diqGeos" between the public services operating in the territory of the municipality of M'diq.

2   Problem Statement

A Geoportal is a web portal used to share and access spatial data and services via the Internet. It can also be defined as a central point of access and sharing of spatial information [5].

A Geoportal is a combination of the Web and GIS to put online 2D or 3D [13] geospatial data (maps, satellite photos, DTM, DEM, point cloud, 3D model …), spatial processing through search tools and spatial analysis [6–8].

The origins of Geoportals go back to the beginning of the Web's development in the 90s when the need for mapping places on the internet has increased. The first geospatial platforms were designed as websites offering the downloading of geographic data files using keywords or through catalogs organized by themes.

Currently, Geoportals have benefited from the development of the web, which has allowed designing interactive interfaces, rich web services in functionalities and a smooth navigation on maps.

Users can use the various web browsers to access Geoportals, through their compliance with HTML protocols, so the user does not need to install heavy and expensive softwares [6].

3   Study Area

Located in northern Morocco, the study area is the municipality of M'diq which is part of the prefecture of M'diq-Fnideq, it is one of the eight provinces/prefectures that make up the Tangier-Tetouan-Alhoceima Region according to the current administrative division.

Fig. 1. Northern Morocco.

Economically: fishing, tourism and trade constitute the main lever of the economy of the M'diq prefecture (Fig. 1).

Demographically, the municipality counts 56,227 inhabitants with an average annual growth rate of 4.34% and the number of households living in the municipality are 13,435 in 2014 with an average annual growth rate of 5.6% according to the 2014 RGPH [9, 10].

Indeed, the municipality of M'diq knows a quick dynamic demographic and urban growth accompanied by strong demand for public services. This growth requires to establish better governance to meet the needs of the population while preserving the natural resources of the site (Fig. 2).

The considerable appeal of the city is also measured by the number of planned projects. As a result, the city of M'diq is getting bigger. This makes indispensable the use of a relevant spatial organization and controlled spatial planning based on innovative and efficient management tools.

Fig. 2. City of M'diq

4 Problem Approach

4.1 Architecture

The Geoportal architecture is a three-tier (3-tier) client/server architecture [5, 11, 12], the first tier is the user interface, provides the data presentation function; the second tier is the processing of GIS transactions on the server side, manages the exchange of information between the data level and the client level; the third tier is the storage of spatial/attribute data (Fig. 3).

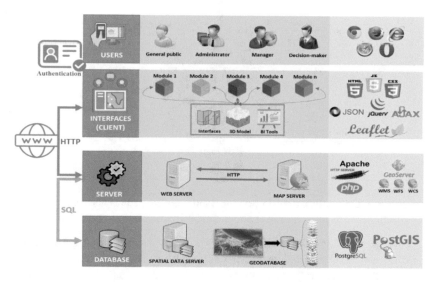

Fig. 3. Architecture of Geoportal M'diqGeos

4.2 Users

Users of the Geoportal can be citizens, private companies, academics, associations, researchers of information and public services.

4.3 Interfaces and Modules

Each stakeholder on the territory of the municipality has its own module equipped with a set of business functionalities according to its needs. The modules are connected to the public services geographic database and can be consulted via the Geoportal interface (Fig. 4).

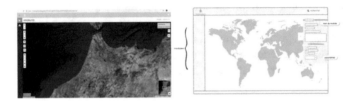

Fig. 4. Main interface of the Geoportal

4.4 Data

For the achievement of this project, we developed specific methodology consisting initially on developing a central geographic database fed by the various public services and organizations (Fig. 5).

Fig. 5. Data exchange between public services and Geodatabase

All the geographical data of the territory are layers constituting the cartographic background of the Geoportal, such as: basic infrastructures (public utility networks: water, sanitation, electricity, telecommunications), urban planning, tourism, cultural heritage, renewable energies (solar cadastre), transport and mobility, agriculture, environment (DTM, DEM, hydrology, land occupation, etc.).

All these layers are superimposed and visualized as interactive digital maps allowing the user to navigate smoothly and search using spatial/attribute (Fig. 6).

Fig. 6. Geodatabase and layers of M'diqGeos Geoportal

4.5   3D Models

3D models offer a three-dimensional representation of the study area, this model is very useful for several applications such as risk management (floods, landslides …), solar cadastre (solar potential of roofs), urban planning implementation of a building), hydrological studies. For this reason, a three dimensional model is a fundamental base for any intelligent Geoportal. In our Geoportal, we use two types of 3D data (Figs. 7 and 8):

• 3D Aerial photogrammetry data and radar interferometry:

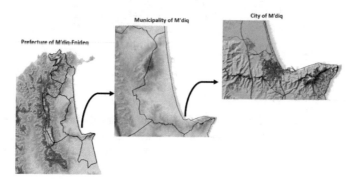

Fig. 7. DTM of the city of M'diq and its periphery

- Terrestrial scans LiDAR data:

Fig. 8. Lidar point cloud 3D scenes samples

4.6   BI Management Tools

In order to become a powerful decision-making tool and a source of useful information for managers and decision-makers, the geoportal is equipped with Business Intelligence tools dedicated to analyzing the stored geodata and process them to produce dashboards (indicators, statistics, maps etc.).

5   Results and Discussions

The figure above shows the different phases of the M'diq Geoportal project (Fig. 9):

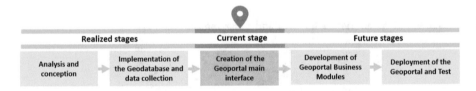

Fig. 9. Stages of realization of Geoportal M'diqGeos

5.1   Analysis and Design

Functionally, we determined the application specifications that constitute the required system, and specify the data and functionalities to be developed.

Technically, the aspects and requirements related to the architecture and the system components were detailed:

- Detailing the architecture, the system modules and the specification of the platform's equipment,
- The nomenclature and the codification of the data in Geodatabase
- Prototypes set of the user interfaces

5.2   Implementation of the Geodatabase and Data Collection

This phase consists of the creation and implementation of a spatial and attribute database of the Geoportal. The spatial database is composed of two categories of data: Basemap Data and Business Data (Fig. 10).

An initial collection of data was carried out: aerial/satellite images, DTM and contour line networks (water, electricity, roads, hydrography), forest, urban planning maps, equipment, cadastral map …

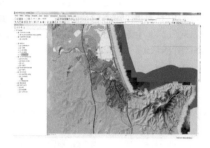

Fig. 10. Geographical data collected

5.3   Geoportal Interfaces

The interface of the Geoportal is very modular and includes on one hand all the necessary functionalities for visualization, navigation, search… on the other hand a collection of public services and organizations modules (Fig. 11).

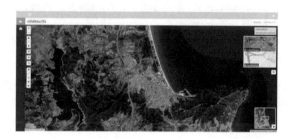

Fig. 11. Main interface of the M'diqGeos

5.4   Geoportal Business Modules

A prototype module rich in functionalities was developed. It's composed of data-ITL tools, update tools, search tools, Business Intelligence tools dedicated to data analysis and reporting (report, statistics, summary sheet ….) (Fig. 12).

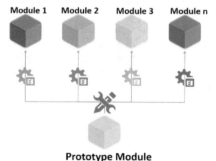

Fig. 12. Adaptation of Geoportal modules

5.5　Geoportal Deployment and Tests

The M'diqGeos is hosted on test server in order to fix the bugs, asses the results and validate the developed modules and functionalities.

After an internal self-evaluation, extern assessment of the Geoportal by users will be carried out to measure the degree of maturity of the system and obtain suggestions for further improvements.

6　Conclusion and Perspectives

In this article, we have presented a solution for implementing a Geoportal dedicated to the sharing and intelligent processing of spatial information. This solution brings innovations to the traditional data sharing portal and is presented mainly at the spatial information level which is a primordial and a complementary source of data.

The main objective of our Geoportal is to provide a unique source of spatial data sharing between public services and organizations for use in their daily tasks.

This Geoportals integrate data from different sources and provide mechanisms and intelligent tools for research, exploitation, analysis … that help in decision-making.

This platform is an example of using Geoportals for smart governance in the implementation of smart city projects.

Future work includes completing the development of the Geoportal, hosting the platform and assessing the user's satisfaction.

Acknowledgments. This work has been supported by MESRSFC and CNRST under the project PR2-OGI-Env, reference PPR2/2016/79.

References

1. Shin, S.-C., Rakhmatullayev, Z.M.: Digital transformation of the public service delivery system in Uzbekistan. In: 2019 21st International Conference on Advanced Communication Technology (ICACT) (2019). https://doi.org/10.23919/icact.2019.8702014
2. Vîlceanu, B., Grecea, C., Herban, S.: Spatial data geoportal for local administration – smart solution for a secure and valuable cultural heritage. J. Geodesy, 7 (2017)
3. Breux, S., Diaz, J.: La ville intelligente: origine, définitions, forces et limites d une expression polysémique Institut national de la recherche scientifique – Centre Urbanisation Culture Société, Montréa (2017)
4. Shahrour, I.: Managing and visualising geospatial data in a user-friendly environment, use of GIS in smart city projects. GIM Int. **28**(4), 21–23 (2018)
5. Becirspahic, L., Karabegovic, A.: Web portals for visualizing and searching spatial data. In: 2015 38th International Convention on Information and Communication Technology, Electronics and Microelectronics (MIPRO) (2015). https://doi.org/10.1109/mipro.2015.7160284
6. Abdalla, R., Esmail, M.: WebGIS techniques and applications. In: WebGIS for Disaster Management and Emergency Response, pp. 45–55. Springer International Publishing, Cham (2019)

7. Tait, M.: Implementing geoportals: applications of distributed GIS. Comput. Environ. Urban Syst. **29**(1), 33–47 (2005). https://doi.org/10.1016/s0198-9715(04)00047-x
8. Nikolaos, A., Kostas, K., Michail, V., Nikolaos, S.: The emerge of semantic geoportals. In: Meersman, R., Tari, Z., Herrero, P. (eds.) On the Move to Meaningful Internet Systems 2005: OTM 2005 Workshops, vol. 3762, pp. 1127–1136. Springer, Heidelberg (2005)
9. Recensement Général de la Population et de l'Habitat de (2014). http://rgphentableaux.hcp.ma
10. Haut Commissariat au Plan. https://www.hcp.ma
11. Jones, J.M., Henry, K., Wood, N., Ng, P., Jamieson, M.: HERA: a dynamic web application for visualizing community exposure to flood hazards based on storm and sea level rise scenarios. Comput. Geosci. **109**, 124–133 (2017). https://doi.org/10.1016/j.cageo.2017.08.012
12. Kan, A., Wang, X., Wu, X.: Geospatial web portal for the Tibetan Plateau ecological safety data services. In: 2011 19th International Conference on Geoinformatics (2011). https://doi.org/10.1109/geoinformatics.2011.5980915
13. Echlouchi, K., Ouardouz, M., Bernoussi, A.S.: Urban solar cadaster: application in North Morocco. In: 2017 International Renewable and Sustainable Energy Conference (IRSEC), Tangier, pp. 1–7 (2017)

Software Defined Networking Based for Improved Wireless Sensor Network

Benazouz Salma[1(✉)], Baddi Youssef[2], and Hasbi Abderrahim[1]

[1] Lab RIME, Mohammadia School of Engineering, Rabat, Morocco
salmabenazouz@gmail.com, ahasbi@gmail.com
[2] Lab STIC, ESTSB, UCD, El Jadida, Morocco
baddi.y@ucd.ac.ma

Abstract. The expansion of Internet of Things (IoT), the promising technology for new intelligent applications, and its fundamental platform Wireless Sensor Networks (WSN), would require a flexible layered architecture to interconnect an increasing number of heterogeneous sensor nodes. The Software Defined Network (SDN) which is an emerging network paradigm that separates control logic from the network device; and brings flexibility and simplicity of network management; has proven extremely useful to cope with WSN defies and to improve its performance. The SDN-based WSN (SDWSNs) have been introduced to reorganize the WSN functionalities according to the SDN model. This paper highlights the importance of adopting SDN in WSN networks, presents the common SDWSN architecture, reviews several contributions on SDWSN architectures and functionalities, analyzes and compares them with focus on aims, architecture design and employed features to ensure improved WSN.

Keywords: Software Defined Networking (SDN) · Wireless Sensor Network (WSN) · Software Defined Wireless Sensor Network (SDWSN)

1 Introduction

The Software Defined Network (SDN) is a network management approach that separates the control plane from the data plane of the network device, and unifies the control planes into external centralized control software "Controller". The SDN controller makes the data routing decisions based on the information received from the network and defines the rules for the incoming flows.

The Internet of Things (IoT) and Wireless Sensor Networks (WSN) are composed of devices that can detect, communicate, process, and route data autonomously, enabling the development of various real applications, such as smart grids, agriculture and smart health, which require the potential deployment of thousands of sensor nodes. However as WSNs grow, networking sensor nodes presents several challenges. One of WSN's main challenges is the constraint of network management and infrastructure flexibility, in addition to resources constraints which include limited resources and storage, communication bandwidth and energy.

The Software Defined Network (SDN) is a promising solution for WSNs enabling separation of control logic from sensor nodes, and thus centralized control and flexible

© Springer Nature Switzerland AG 2020
M. Ezziyyani (Ed.): AI2SD 2019, LNNS 92, pp. 246–258, 2020.
https://doi.org/10.1007/978-3-030-33103-0_25

management of the entire WSN network. Software-defined wireless sensor networks (SDWSN) have been introduced as an approach to SDN application in WSN.

This paper highlights various recent contributions on SDN-based architectures for WSNs in better detail, while focusing on the advantages that SDN brings to WSN in the matter of design, management and functionalities. This paper also focuses on analyzing and comparing several aspects that must be considered in the design of SDWSN architecture such as aims, components, controller composition, and designed and employed features in SDWSN architectures, in order to enable addressing challenges of WSN.

This paper is organized as follows. The first section presents a survey of SDN technology, the WSN, and the challenges faced by WSN. The following section discusses the WSN's SDN approach as an effective solution to many WSN-related challenges, leading to improvement of its performances. The third section presents the general SDWSN architecture as applied by various researchers, and reviewed several contributions on SDWSN architectures and functionalities, in particular, Sensor OpenFlow, Soft-WSN, SDN-WISE, TinySDN and SDWSN-RL. The last section underlines an analysis and comparison of these architectures with focus on aims, architecture design and employed features to ensure improved WSN.

2   Background and Terminology

2.1   SDN

The Software Defined Network (SDN) is a new concept based on the idea of separating the control plane from the data plane, and moving this decision-making part of the equipments to a centralized control plane. This plan acts as a mediator between the applications and the equipments, allowing an abstraction and a programmability of the network [1].

The SDN comes with a very structured architecture in three layers (Fig. 1) that communicate with each other via open programming interfaces APIs [2].

The application or management plane consists of SDN software applications that implement network services (such as routing, security, quality of service). These applications have an abstract view of the entire network that ensures their interaction in real time with the network [2].

The control plane is formed by the SDN controllers. The SDN controller is a logical entity centralized on a server. It retrieves information about the devices network for controlling abstraction. It defines the rules for incoming flows from the data plane according to the requirements of the application plane, and programs the routing devices via APIs [2].

The data plane is responsible for the transfer of data packets. Forwarding devices act on instructions in flow tables assigned to them by the controller to handle incoming packets. A flow represents a sequence of packets with values serving as a filter criteria and a set of actions to execute [2].

The Northbound API facilitates the coordination of application developers with the network operating system (NOS) associated with the controller and enables efficient network automation [2]. The Southbound API provides a communication protocol that helps the controller program the routing devices [2].

The Software Defined Network (SDN) brings innovation, flexibility and simplicity to network management and that in various networks such as data centers, WAN networks, etc.

The growing development of the Internet of Things (IoT) paradigm is leading to the implementation of SDN technology in the IoT network to improve its efficiency and intelligence. The Wireless Sensor Network WSN represents a fundamental platform for the IoT network. This leads to study primordially the integration of SDN into WSN, to meet the requirements of IoT and to solve the inherent problems in WSN [3].

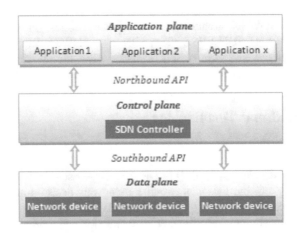

Fig. 1. The basic SDN framework.

2.2   WSN

A Wireless Sensor Network (WSN) is a network of multiple sensors spatially dispersed to acquire physical data; calculate information from the collected values, and route them through a multi-hop routing to the sink node (gateway). The sink node transmits this data via the internet or satellite to the network user (task manager) to analyze the data and make decisions. WSNs are mostly exploited for tracking and monitoring applications, such humidity, temperature, etc. [4].

The WSN is one of the applications covered by IEEE 802.15.4, a communication protocol designed for Low-Rate Wireless Personal Area Network (LRWPAN). It is used by many implementations based on IP or proprietary protocols, such as 6LoW-PAN and ZigBee [5].

The ZigBee/IEEE 802.15.4 architecture is the most popular architecture used for WSN networks. IEEE 802.15.4 defines low layers, the Physical Layer (PHY) that contains the Radio Transmitter/Receiver, and over it Medium Access Control Layer (MAC), which defines the data processing format and manages how a ZigBee node will "speak" in the network. The ZigBee standard defines the communication of upper layer protocols; the Network Layer defines the rules for establishing a network of nodes, the association and interconnection of nodes, as well as the format of the frames that will be exchanged. The application layer will determine how all lower levels are used for a given application [5].

2.3   WSN Challenges

As WSN grow, the networking of sensor nodes presents several challenges. Some of these key challenges are the difficulty of network management, the rigidity to policy changes, the realization of a heterogeneous network of nodes, and resources constraints such as limited processing and storage capacity, energy and communication bandwidth [4].

In fact, in large and highly active WSN deployments, low data throughput between nodes can cause congestion issues, which can affect the overall throughput and latency of the network [5].

The energy constraint is in particular the most important challenge in the development of the WSN, because it determines the lifetime of the network. The sensor nodes consume a lot of power during detection, communication and data processing and operate with a limited battery [6].

2.4   SDN Approach of WSN (SDWSN)

Many studies have identified that SDN addresses most of the challenges faced by WSNs, and presents a heterogeneous integration model [3]. The application of SDN in WSN gives rise to a Software-Defined Wireless Sensor Network SDWSN.

The abstraction of SDN makes it possible to maintain a global view of the underlying network to the applications, thus the efficiency and topological organization of the WSN given its significant scalability with the advent of IoT [5].

SDN enables flexible network management, and ability to program sensor nodes and change the transmission path in real time by inserting diverse packet forwarding rules, to provide the best routing policies through controller.

Because of the decoupling virtue in SDN, most energy-intensive functions such as routing, traffic management and processing are removed from the physical node, and managed at the controller level that has enough power resources or at the application level. Thus, a considerable amount of energy is saved to potentially extend network life [5].

The SDN approach in WSN promotes efficiency and sustainability in WSNs as they grow.

3   State of Art of SDWSN Architecture

The SDN approach of Wireless Sensor Networks consists in reorganizing the various WSN functionalities according to the three logical planes of the SDN model: application, control and data [5]. The development of the SDWSN architecture is still in its infancy. However, different implementations of SDWSN architectures conforming to the fundamental principles of the SDN, notably the decoupling, have been realized and proposed by the researchers.

The Fig. 2 represents the general SDWSN architecture as applied by various researchers.

3.1   Implementation of SDWSN Architecture

The data plane comprehends the sensor nodes including hardware and software components. The hardware consists of a radio, a power supply unit, and a sensing unit. The hardware components are on the physical layer (PHY) that associates with Media Access Control layer (MAC) to fulfill the IEEE 802.15.4 functionalities specified for LR-WPAN. The software part of the sensor node plays a vital role in the processing of detected data and routing functionalities. It typically includes a data generation module which is added on each sensor node in SDWSN for traffic generation, since the nodes in WSN generate data packets on their own that are subsequently transmitted to the processor. SDWSN model offers an in-network processing module that performs data aggregation at the sensor node and data fusion at the sink node, to reduce data redundancy issue in WSN. If processing is not needed, then it simply forwards it to the flow table. The flow table stores the rules prescribed by the controller. The abstraction layer provides an interface to communicate with the controller [5, 6].

The control plane consists of one (or possibly several) controller centralizing the control logic and all the intelligence of the network. The core features of the SDWSN controller are flow rules generation, mapping functions, and programming interfaces. The controller starts creating the flow rules based on different aspects of the mapping information used for routing purposes. The controller sends out a monitoring message to the sensor nodes, to acquire information about the state and the topology of the network and keep it up to date, such as energy level, distance to the sink node, neighbor list and link status parameters. The acquired information is stored in the mapping information module. The mapping function is the component that processes the monitoring data and creates a network topology view [5, 6].

The application plane houses applications such as network management and functionalities in the WSN using information from the control plane [6].

The ongoing research on architecture conception and network management in the SDWSN model has made considerable progress, but in its beginning. SDWSN architectures vary slightly in their implementation. The functions of the MAC and PHY layers on the forwarding node remain in WSN, however disparities in the implementation of data processing and also of the control logic vary in various research studies; for example, some place the control logic at the sink node and others have it at a higher level.

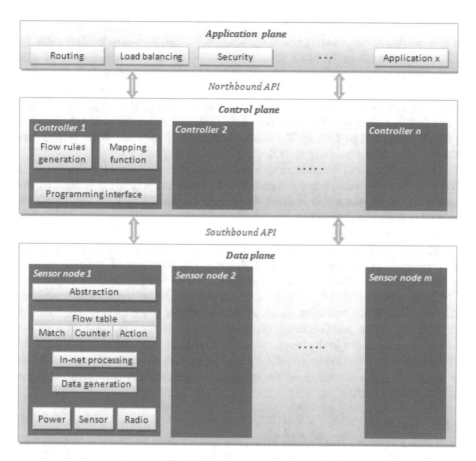

Fig. 2. SDWSN architecture.

3.2   SDWSN Architectures

Several existing architectures for implementing SDN for WSN will be discussed below in this article; however, there are other solutions that are also available.

3.2.1   Sensor OpenFlow

One of the first works that synergizes SDN and WSN was published in IEEE's communication letters in 2012 by Luo et al. They proposed SDWSN architecture with a standard communication protocol between the control and data planes, Sensor Open-Flow (SOF) [7].

OpenFlow was designed as a wired protocol for enterprises and carrier networks, used to connect and transmit control messages between a controller and a high-speed switch. The Sensor OpenFlow (SOF) is similarly defined to address the disparities and challenges faced by WSNs. The idea is to manipulate a programmable flow table on each sensor via SOF.

Unlike address-centric OpenFlow networks where the mappings in the flow table entries are based on a specific IP addressing, WSN are data-centric. To handle the addressing schemes in WSN for flow creation, the OpenFlow-compatible SOF flow entries (Match) are redefined as Class-1 entry (compact network-unique addresses), where the data forwarding is centered on the node ID such as ZigBee 16-bit network addresses; and a Class-2 entry (concatenated attribute-value pairs CAV), where the data forwarding is centered on the value (i.e., if the sensed data exceeds a certain threshold value, the data will be routed). The second solution is to augment WSN with IP, two standard IP stacks are recommended: uIP/uIPv6 and Blip.

On the other hand, the additional control traffic carried by the SOF channel in the WSN band (i.e., in band) can overload the resource-limited underlying WSN. Then the latency of the control traffic is much larger than in OpenFlow, since the OpenFlow channel is usually out of band.

Indeed, the control traffic is mainly composed of two types of control messages: (1) Packet-in sent by a sensor to a controller to obtain instructions for a packet that does not correspond to any flow entry; and (2) Packet-out or flow-mod which is the response of the controller to the requesting sensor.

The resulting control traffic can be significant because of the dynamic and irregularity of the WSN traffic; since an event can trigger a set of flow configuration requests, associated with expiring timers, from multiple sensors simultaneously. To curb the control traffic in the Sensor OpenFlow solution, one packet-in is sending for the first table-miss and the subsequent Packet-in matching the same address destination as the first one are deleted, until the corresponding Packet-out is received or a predefined timeout occurs.

3.2.2   Soft-WSN

Bera et al. proposed the Soft-WSN [8] architecture which adopts a real-time device and network topology management. Therefore, they designed a software-defined controller having two management entities, device manager and topology manager, to improve the QoS of the network.

The device manager allows controlling the use of the sensors according to the application-specific requirements. It schedule in real-time the sensing tasks by activating the required sensor within a sensor node composed of multiple sensors, the sensing delay, and the active-sleep state of the sensor node.

The topology manager is responsible for managing forwarding rule and network-connectivity depending on the requirements such as security and QoS. It consists of node-specific management that allows to modify the forwarding logic of a particular nodes (i.e., change-dest, do-not-send, and do-not-receive); and network-specific management that defines a forwarding rules for all nodes in the network (i.e., to drop a particular node's message (drop-from) or do-not-forward to a particular node). The required instructions are defined in the Action field of the described packet format.

Bera et al. have achieved experimental results on a test bench based on real hardware. The packet delivery ratio increases (i.e., more data packets reach to the destination) since the routing path is maintained by the controller. Soft-WSN reduces the replicated data packet due to the definition of forwarding rule, thus the data message overhead. However, the number of control messages is higher, as the node communicates with the controller each time it receives a new packet. Soft-WSN minimizes the energy consumption by scheduling sensor nodes tasks. These results show that the performance of the network can be improved using Soft-WSN compared to the traditional WSN.

3.2.3 SDN-Wise

Galluccio et al. introduced a stateful SDN solution for WSN; SDN-WISE [9]. The basic concepts of the SDN-WISE protocol are derived by OpenFlow, the sensor nodes manage a WISE flow table, whose entries placed by the controller, specify the matching rules and actions. The main distinguishing feature as compared to traditional OpenFlow is that the SDN-WISE is stateful. A memory buffer is reserved for the current state value within each SDN-WISE node in its WISE States Array. The state can be modified by executing the actions. Also, given the broadcast nature of the wireless medium, each sensor node has to select and further process only the packets which the Next Hop ID field is listed in their Accepted IDs Array. The packet classification also differs compared to Openflow since WSN are data-centric. In SDN-WISE Matching Rules may consider any portion of the packet and the current state. The action in WISE Flow Table entry can impact the packet as well as the current state (i.e., supported Actions are Forward to, Drop, Modify, Send to INPP, Turn off radio).

SDN-WISE is energy efficient. It supports data aggregation and the control of duty cycle; nodes can be turned to the OFF state periodically.

Furthermore, the sensor nodes are enabled to execute stateful programmable operations without requesting the controller, if local strategies are to be applied. The goal is to reduce interactions with the controller and achieve system efficiency. For example, for conditional packet forwarding from one node to another, or for QoS support, by storing the congestion level as state information or determining the priority of flows in a congested node.

The SDN-WISE architecture is composed of sensor node, sink node, and controller. SDN-WISE also contain tools for running a real Controller in an OMNeT++ simulated network.

Each node in SDN-WISE encloses PHY and MAC layers (IEEE 802.15.4), For warding layer (FWD) and In-Network Packet Processing (INPP) layer. It use as well an appropriate protocol run by the Topology Discovery layer (TD) to generate local topology information and indicate the best next hop towards one of the sinks to contact the controller.

The Topology Management (TM) layer in the WISE-Visor Controller built a table with incoming TD packets (battery level and the distance to the nearest sink) and the WISE Neighbors list) in order to always have an updated view of the local topology and the current network status.

3.2.4 TinySDN

Tiny-SDN [10] is a flow-ID-based SDWSN approach enabling the implementation of multiple controllers, thus decreasing the latency of the network due to in-band control traffic of WSN. It is the SDN implementation for the WSN operating system, TinyOS. The Tiny-SDN architecture contains: SDN-enabled sensor nodes and SDN-controller node (or nodes) which are connected through multi-hop wireless communication.

TinySDN utilizes the collection tree protocol (CTP) as underlying protocol. CTP is hardware-independent; it adopts the link quality metric (ETX) between single-hop neighbors based on the packet delivery amount. The route with the lowest ETX sum is used to send data to a sink node (root), attached to the controller trough a serial/USB connection. Therefore, a collection tree routing is built for each sink node thus providing multiple SDN controllers.

The SDN-enabled sensor node in data plane runs on sensor mote and has three main components: TinyOS Application is the equivalent to end-user device which generate data packets; TinySdnP is responsible for performing flow table matches, related actions (i.e., drop and forward) and updates. And ActiveMessageC which manages and provides programming interfaces to handle radio module of the nodes (such as link quality estimation and control/data packets forwarding).

Each SDN-enabled sensor node have to recognize and measure link quality with its neighbors, and assign to an SDN controller node to send network topology information and possibly flow setup requests through the CTP route.

The SDN controller node maintains two modules: Sensor mote module which is responsible for communicating with SDN-enabled sensor node using ActiveMessageC, and Controller server module which hosts the controller applications and manages the network traffic flow and the topology information to keep a network topology view. The sensor mote (sink node) uses as well SerialActiveMessageC to communicate with the controller server, and TinySdnControllerC to adapt messages and manages this communication.

3.2.5   SDWSN-RL/Cognitive SDWSN

Huang et al. propose SOF-based Cognitive SDWSN [11] prototype providing energy saving with guaranteed QoS and self-adaptability of WSNs. They design reinforcement learning (RL) based mechanism embedded in the control plane.

The application plane of SDWSN prototype includes an information QoS setting module to meet the QoS requirements of applications.

The data plane is abstracted into a weighted directed graph. The link weight refers to the link bandwidth utilization (BWU) that indicates the distribution status of flows. It handles the information processing and employs Over-the-Air Programming (OTAP) technique for dynamic configuration to improve programmability and resource reutilization.

The controller in the control plane manages data fusion and routing. It includes Cognitive Information Middleware (CIM) that evaluates current value-data flow using an ARMA (autoregressive moving average) model to exploit possibly value redundancy fusion, and builds maps of flow distribution and network interconnection. CIM utilizes these results to produce optimal routing decisions and define policies to be deployed in the flow table (Value and Path matching are designed) for a lightweight implementation of data plane. This leads to decrease data-gathering traffic in the data plane and control overhead in the control plane.

SDWSN-RL achieve energy saving by mainly using value-redundancy filtering, to control the traffic forwarding adaptively (i.e., configuring high medium access delay for low-value flow), and implement value-redundancy fusion. It use as well load-balance routing, by analyzing the flow distribution and adjusting the path weights (i.e., a link with higher bandwidth utilization (BWU) will be configured with a larger contention window (CW)).

4 Discussion and Comparison Study

The objectives and requirements considered by SDWSN architectures design and implementation are: enabling the SDN implementation in WSNs taking into account the WSN constraints and problems, apparently energy, resources, value redundancy, data-centric nature of WSN typically, difficulty to manage, and rigidity to policy changes.

We focus in Table 1 on several aspects that must be considered in the design of SDWSN architecture such as aims, components, controller composition and implementation, cross-plane communication protocol. We analyze and compare employed features designed and implemented in SDWSN architectures reviewed above, in order to allow addressing challenges of traditional WSN.

Table 1. Comparison of various SDWSN architectures.

Architecture	Sensor Openflow	Soft-WSN	SDN-WISE	TinySDN	SDWSN-RL
Focus	Design SDWSN SB API Protocol Flexibility Flow identification Limit control traffic	Control of device (sensor scheduling) and topology Ensure QoS Flexibility Management simplicity	Program sensor node as finite state machine Energy management Control duty cycle Data aggregation Flexibility Reduce control signaling	Use of multiple SDN controllers TinyOS-based SDN Hardware independence (CTP) Resource management	Energy efficiency Load-balance routing Manage value-redundancy Limit control overhead Adaptability Guarantee QoS
Components	Controller – Sensor Nodes	Sensor nodes – AP – R/Sw – Controller – Data server	Controller (s) – Sink – Sensor nodes	SDN-enabled sensor nodes Sink node serially attached to SDN-controller node(s)	Controller(s) – Sensor nodes (clusters)
Controller composition		Topology manager Device manager	WISE Visor: Topology manager	Sensor mote Controller server	Cognitive Information Middleware

(continued)

Table 1. (*continued*)

Architecture	Sensor Openflow	Soft-WSN	SDN-WISE	TinySDN	SDWSN-RL
Cross-plane protocol	SOF	IEEE 802.11		Multi-hop Wireless Communication	SOF
Controller	Central or Distributed	Central	Distributed	Distributed	Distributed
Derived by Openlflow	✓		✓	✓	✓
In-network Processing	✓		✓		✓
Redefining flow entries	✓		✓	✓	✓
Curbing control traffic	✓	✗	✓	✓	✓
Reducing message overhead	✓	✓ (Data)	✓	✓	✓ (Data/Control)
Supporting duty cycle		✓	✓		
Saving energy	✓	✓	✓	✓	✓

The analyzed features concern, in one hand, the implementation of in-network data processing on sensor nodes, like data aggregation which refers to summarize data of multiple sensors to reduce redundancy, or data fusion which refers to associate the information combined from multiple sensors.

On the other hand, we review the SDWSN contributions which redefine flow entries, to adapt to the data-centric addressing schemes in WSN, as the main interest in WSN is based on data exploring rather than source addresses. SDWSN-RL use value/path matching in flow table. TinySDN employs flow ID based approach. SDN-WISE design further operators for packet classification and consider the current state in matching rules. Sensor OpenFlow develop a class 1 and class 2 entries to forward value centric data.

Moreover, we review the suggested solutions for control traffic curbing, since the decoupled structure in SDN result to additional in-band control traffic between sensor nodes and controller(s), such as deploying programmable SDN-WISE sensor nodes enabled to execute stateful operations without requesting the controller, controlling table-miss in Sensor OpenFlow, designing optimal rules in SDWSN-RL flow tables, and implementing multiple controllers in TinySDN.

We review as well the minimization of message overhead, comprising the overhead for control packets and data packets. Sensor OpenFlow and SDN-WISE utilize solutions to curb control traffic as mentioned before, thereby must reduce control message overhead. For Soft-WSN, only the data overhead is reduced by controlling data packets replicas using unicast or multicast forwarding between sensor neighbors instead of broadcast [8]. For SDWSN-RL, both data and control overhead must be minimized due to value-redundancy control and flow distribution adjusting.

The commonness of the OpenFlow protocol in SDN applications seems influencing even the SDWSN models such SDN-WISE, SDWSN-RL, Sensor OpenFlow and TinySDN.

Additionally, saving energy is as well evaluated being the main defy in WSN. It is achieved by centralized control first, and by the implementation of various mechanisms like: in-network data processing, controlling data redundancy and limiting control traffic and message overhead, in order to conserve network resources (like in Sensor OpenFlow, TinySDN, SDWSN-RL and Soft-WSN), supporting duty cycle (like in Soft-WSN and SDN-WISE), plus programming stateful operations at the SDN-WISE sensor nodes (finite state).

5 Conclusion

This survey provides an overview of SDN and WSNs, and introduces the concept of Software-Defined WSNs (SDWSNs) taking into account the challenges in WSNs, particularly, the energy and resources constraints, the difficulty to manage and to change policies. Implementing SDN approaches reveals improving the performance of WSN compared to traditional approaches.

We surveyed and discussed some existing SDWSN architectures, with focus on design (such aims, components, controller composition, etc.) and employed features to cope with challenges. The reviewed SDWSN architectures that are yet to be real implemented and fully realized, attempted to handle these challenges by allowing an abstraction and a programmability of the network and proposing various designs and solutions as studied above the article.

However, it still remain many defies. In fact, the control-data decoupled structure and logic centralization of SDN may result in excessive cross-plane control traffic overhead, transmission delay, and thus require conception study of network and sensor node.

It is important to include practical part of Software Defined Wireless Sensor Network for future work. We will deploy a virtual network using SDN emulator Mininet, sensors network simulation tool for Contiki Cooja, ONOS for controller, and evaluate its performance.

References

1. Joshi, S., Dutta, A., Chavan, N., Dhus, G., Bharsakle, P.: A survey on software defined networking in IoT, vol. 6, no. 2, p. 4
2. Khan, S., Ali, M., Sher, N., Asim, Y., Naeem, W., Kamran, M.: Software-defined networks (SDNs) and internet of things (IoTs): a qualitative prediction for 2020. Int. J. Adv. Comput. Sci. Appl. **7**(11) (2016)

3. Pritchard, S.W., Hancke, G.P., Abu-Mahfouz, A.M.: Security in software-defined wireless sensor networks: threats, challenges and potential solutions. In: 2017 IEEE 15th International Conference on Industrial Informatics (INDIN), Emden, pp. 168–173 (2017)
4. Mallick, C., Satpathy, S.: Challenges and design goals of wireless sensor networks: a sate-of-the-art review. Int. J. Comput. Appl. **179**(28), 42–47 (2018)
5. Kobo, H.I., Abu-Mahfouz, A.M., Hancke, G.P.: A survey on software-defined wireless sensor networks: challenges and design requirements. IEEE Access **5**, 1872–1899 (2017)
6. Ali Hassan, M.: Software defined networking for wireless sensor networks: a survey. Adv. Wirel. Commun. Netw. **3**(2), 10 (2017)
7. Luo, T., Tan, H.-P., Quek, T.Q.S.: Sensor OpenFlow: enabling software-defined wireless sensor networks. IEEE Commun. Lett. **16**(11), 1896–1899 (2012)
8. Bera, S., Misra, S., Roy, S.K., Obaidat, M.S.: Soft-WSN: software-defined WSN management system for IoT applications. IEEE Syst. J. **12**(3), 2074–2081 (2018)
9. Galluccio, L., Milardo, S., Morabito, G., Palazzo, S.: SDN-WISE: design, prototyping and experimentation of a stateful SDN solution for WIreless SEnsor networks. In: 2015 IEEE Conference on Computer Communications (INFOCOM), Kowloon, Hong Kong, pp. 513–521 (2015)
10. de Oliveira, B.T., Margi, C.B.: TinySDN: enabling TinyOS to software-defined wireless sensor networks, p. 8 (2016)
11. Huang, R., Chu, X., Zhang, J., Hu, Y.H.: Energy-efficient monitoring in software defined wireless sensor networks using reinforcement learning: a prototype. Int. J. Distrib. Sens. Netw. **11**, 1–12 (2015)

SWCA Secure Weighted Clusterhead Election in Mobile Ad Hoc Network

Meriem Ait Rahou[✉] and Abderrahim Hasbi

Mohammadia School of Engineers, Mohamed 5 University, Rabat, Morocco
mariamaitrahou@research.emi.ac.ma, hasbi@emi.ac.ma

Abstract. Mobile ad hoc networks (Manets) consist of a large number of mobile nodes communicating in a network using radio signals. Clustering is one of the techniques used to manage data exchange amongst interacting nodes. Each group of nodes called cluster and has one or more elected Cluster head, where all cluster heads are interconnected for forming a communication backbone to transmit data. Moreover, cluster heads should be capable of sustaining communication with limited energy sources for longer period of time. Misbehaving nodes and cluster heads can drain and reduce the total life span of the network. In this context, selecting the appropriate cluster heads with trusted information becomes critical for the overall performance. In this paper, we analyze the existing trust based clustering solutions, and highlight their advantages and drawbacks. Besides, we propose an efficient trust based cluster head selection.

Keywords: Manet · Clustering · Trust · Cluster head

1 Introduction

A mobile ad hoc network (MANET) is a self-configured, multi-hop and dynamic network of mobile nodes connected by wireless link and without any additional infrastructure or centralized management. It has numerous distinguishing particularities, such as: (1) dynamic topology: nodes may move randomly with different speeds; (2) autonomous: every mobile node act as router as well as host; (3) multi-hop routing: communication with two nodes which are out of transmission range of each other, then packets are forwarded via intermediates nodes; (4) energy constraint: every entity in Manets rely on its own battery;..Etc.

Actually, this topology of network is introduced in many domains applications [1] to name but a few, emergency search, rescue operations, meeting or conventions, electronic email, file transfer, and military application.

In mobile ad hoc network, to solve capacity and scalability problems and to have a good utilization of nodes resources, efficient management and prolonged network lifetime, one prominent way of network management have been proposed in the literature namely clustering. This approach is a process that divides the network into interconnected substructures, called clusters. In a clustering scheme, all the mobile nodes are grouped into different geographically distributed groups [2]. Within every single cluster, there are three kinds of nodes: ordinary node, gateway node and cluster

© Springer Nature Switzerland AG 2020
M. Ezziyani (Ed.): AI2SD 2019, LNNS 92, pp. 259–265, 2020.
https://doi.org/10.1007/978-3-030-33103-0_26

head node CH. The first one, forward data to the cluster head and cannot communicate directly with nodes of another clusters. The second one connects cluster with other clusters. The third one represents the particular node. It manages all these communications, looks after cluster maintenances and updates the routing table. It's the main and leader node in a cluster. Therefore, many researchers paid attention to reliable cluster head selection techniques in mobile ad hoc network.

Those techniques are classified by [3] under four categories as (1) Arbitrary metrics-based which use random values such as id nodes. (2) Specifics metrics-based in which CH election depends on a particular metric such as energy, topology, mobility, security, load balancing and so on. (3) Without metrics-based in which CH is chosen without resorting to any selection metrics such as Passive clustering [4] and finally (4) Combined metrics-based in which CH is elected according on many parameters at the same time including energy, mobility, node degree, communication range, ability to handle nodes and so forth. Thus, the last category is considered as the distinguished one.

In the combined metrics category, Weight clustering algorithm (WCA) [5] is one of the most important approach proposed for MANETs. It calculates weights of nodes and selects the node with the minimum one as cluster head. WCA minimize communications overhead by invoking CH election procedure only if nodes mobility changes or even if current CH is unable to cover all nodes and not in a periodic manner. Moreover, it avoids cluster heads overload by predefining a threshold that indicates the number of nodes each CH can ideally support. Although all this advantages, WCA haven't taken security features into account.

Security in mobile ad hoc network can be categorized into three groups as in [6], trust based, cryptography based and hybrid schemes. Each of these categories protects the clustering process against special type of attacks. For instance, cryptography based clustering techniques defends the outsider attacks and the trust based clustering techniques defends the insider attacks and prevent the election of malicious or compromised node as CHs or other cluster components. In this paper, we highlight the most prominent trust based clustering techniques in manets, and try to combined it with the WCA in order to select the appropriate cluster head with the minimum weight and in a secure manner in the same time.

The remainder of this paper is organized as follows: Sect. 2 introduces the related work. Section 3 presents our proposed cluster head selection algorithm in manets. Section 3 provides a discussion and analyzes and finally Sect. 4 concludes the paper.

2 Related Works

In existing algorithms in mobile ad hoc networks, some of them consider only trust of nodes in cluster head election process, while only a few of them consider both trust and weight clustering algorithm WCA. Some proposed algorithms in the literature take into account trust value as unique metric to form clusters and elects cluster head. In these, only trusted nodes could participate in network communication. A node is considered as trusted according to the evaluation of its interactions with neighbors. [6] Cites that trust-based clustering algorithms are classified into two types of algorithms: pure and hybrid.

To name but a few, the trust evaluation metrics are the type of data forwarded, data delivery rate, recommendation by CH, block list value and Number of packets forwarded, dropped, misrouted and falsely injected.

A Secure Clustering Algorithm in Mobile Ad Hoc Networks SCAR. It is a secure clustering algorithm [7] based on combined metrics including node's reputation value, the node's degree and the relative mobility. It's divided into two steps cluster formation and cluster reestablishment.

In the first step, every node broadcasts Hello message periodically to its neighbors including the initial weight information for connectivity. According to the number of hello messages received, degree, mobility, power transmission and node's reputation are updated for each node. The node with highest weight and highest reputation is elected as cluster head and nodes are added to join in its cluster. If a node hasn't received the cluster head's message during a period, it becomes an isolate cluster head which has no cluster member.

In the cluster reestablishment step, a reputation threshold is given to every node's role i.e. cluster head, gateway and cluster member. If a node's reputation is higher than its reputation threshold this node is considered as suspicious. Then if this node is cluster head or gateway, the algorithm searches its neighbor's reputation. If it is higher than this suspicious node's, cancel the suspicious node's role of cluster head, and elect this node as cluster head. Else, keep the suspicious node's role. If the suspicious node is cluster member, put it into the black list and isolate from network.

Security and stability of network can be effectively performed by SCAR, however, it's occurs an overhead in the system and noting that it's try to secure clustering mechanism against wormhole attack in ad-hoc networks (communication between CHs) but after forming clusters, not during the election procedure of CHs [8].

Improved Weight Based Clustering Algorithm IWCA. This algorithm [9] consists of excluding malicious node from clustering process in order to improve network stability and reliability. It is divided into two phases, malicious node detection phase and clustering phase.

In the first one, each node observes and exchanges the nodes behaviors within its radio transmission range. A node's behavior is described in terms of the ratio of the amount of this behavior over the total amount of packets that the node has received, such as packet drop rate (PDR), packet misroute rate (PMR) and packet modified rate (PMOR).

The observed behaviors are collected in a dataset by the behavioral data collection module. When a node needs to summarize its observation of misbehaving neighbors, it will calculate the rate of abnormal behaviors over the overall behaviors it has observed as follows:

Packet drop rate PDR = Number of Packet Dropped/Total Number of Incoming Packets.
Packet misroute rate PMIR = Number of Packet Misrouted/Total Number of Incoming Packets.

Packet modification rate PMOR = Number of Packet Modified/Total Number of Incoming Packets.

Once the calculation is done, for every node, a threshold is checking for PDR, PMIR and PMOR in order to determine if it's malicious or normal. The weight is then calculated. The node having the minimum weight with normal behavior is stated as cluster head and all nodes within its transmission range with normal behavior are stated as cluster member. Thus, malicious nodes are removed and exclude from cluster formation after having calculated their weights which drain nodes capabilities.

A Weighted Clustering Algorithm Trust Model WCMT. This approach [10], elects the cluster head according to the weight and trust value of the node simultaneously. Basically the weight of node is calculated based on WCA by combining various parameters such as node degree, power of node and mobility. The trust (1) of the node A on another node B is calculated as following:

$$T_{AB} = W_1(PDR) + W_2(PMR) + W_3(PFR) + W_4(PAR) \tag{1}$$

Where $W_1 + W_2 + W_3 + W_4 = 1$ and Packet Dropping Ratio (PDR) is taken as cumulative sum of number of dropped packets per unit time.

Packet Misrouting Ratio (PMR) is taken as cumulative sum of number of packets Misrouted per unit time.

Packet falsely injected Ratio (PFR) is taken as cumulative sum of number of packets falsely introduced into the network per unit time.

Packet Altering Ratio (PAR) is taken as cumulative sum of quantity of packets altered by the node per unit time.

At First, node sent its ID and trust value to its neighboring nodes with REQ/REP flag. When participating nodes identify their neighboring nodes, they share data regarding one-hop neighboring nodes. Then, WCA is invoked and the node with less weight is selected. If the trust value of this node is lowest than the threshold value, then it's chosen as cluster head. The nodes with the trust value highest than the threshold are neglected and eliminated from network. The process repeats till every node in the network is either assigned as CH or as a member.

Clustering Algorithm for Security in Ad Hoc Networks CASAN. This algorithm [11] presents a clustering algorithm to elect the suited node as CH by taking into account the node trust level, the mobility, distance to neighbors, remaining energy and connectivity degree. For this, each node broadcasts a hello message including TTL (time to live) including its identification and its index mobility. Then, nodes with trust level less than a threshold execute the non-trustworthy nodes procedure while others execute the trustworthy nodes procedure. In this last, each node computes its connectivity, broadcasts its metric components and uses the received one to computes its weight as well as its neighbors weights. If the node has the minimum weight compared to its neighbors, it declared itself cluster head by sending the role message CHMSG to its one-hop neighbors. Otherwise, it launches a timer and waits for role messages from its neighbors with lower weights. If it receives many role message CHMSG, it will join the CH with the lowest weight and then broadcast a role message to its one-hop neighbors to affirm its role as an ordinary node.

A Secure Clustering in MANET Through Direct Trust Evaluation Technique. TBCA is one of the important proposed algorithms [12]. In this, cluster head is elected once the trusted nodes are designed. At first trust value of all nodes is calculated within the formula (1). The values are broadcasted through hello message to all neighbors. If trust value of a node is highest than the trust total threshold value, this one is elected as cluster head.

Then if this value is highest than the predefined threshold value, the node is considered as trusted.

Voting Based Clustering Algorithm. VCA [13] is another trust based clustering scheme which evaluates stability of node through computing the neighbor change ratio and the residual power of nodes. In this scheme, each node votes other nodes only if the node is the most trustful one among its neighbor nodes. Votes are propagated only to one hop neighbors and they are not forwarded by other nodes. For clustering, at first each node computes its stability. Then it computes the trust of node with the respect to its neighbors. And finally, each node votes its neighbors according to the voting algorithm. Choose the largest one as the CH but if the number of votes is the same, choose the best stability as CH. If the number of votes and stability are the same, choose the smallest ID as CH.

Secured Weight-Based Clustering Algorithm or SCA. This algorithm [14] elects CH according to their weight computed by combining a set of parameters such as stability, battery, degree and etc. To create or maintain clustering architecture, first the discovery stage should be done which information about the neighborhood should be retrieved. For this purpose, nodes desiring CH send CH_ready beacons to their D hops radius. Then nodes receiving this beacon Estimate a trust value and send it back. After a discovery period, nodes having initiated this operation can derive from the received responses information such as degree, stability trust value. Then each node adds to the previous parameters the state of its battery and the max value and combines them to computes the global weight. After nodes choose the cluster head which had the maximum weight, each newly elected cluster head needs to discover other to construct a virtual backbone for inter-cluster communication. Thus every new elected cluster head broadcasts a discovery request over the network. Cluster heads receiving this request register the certificate of the new cluster head their certificate.

3   The Proposed Algorithm SWCA

In this paper we present our algorithm namely SWCA which based on combining trust parameter to the well known algorithm WCA. It's divided into many steps: Behavior classification, weight calculation, weights comparison and then cluster head electing. The first one consists of classifying every node behavior based on number of packets received, dropped, misroutes, or modified.

In SWCA, if the node is malicious, it will be automatically excluded from clustering process with no need of its weight calculation. Thus SWCA minimize network overhead and forming clusters based on trusted nodes.

(1) Behavior Classification
   For each node (i)
      (1) calculates:

   - No. of packet received PR
   - No. of packet drop PDR
   - No. of packet misroute PMIS
   - No. of packet modified PMD

      (2) Check threshold value of PR, PDR, PMIS, PMD
      (3) Assign behavior to node according threshold value malicious or normal (If the value is > threshold value then the node is malicious)
(2) If the node (i) is malicious then (i) is eliminated from clustering process.
(3) Else If the node is normal
   For each node (i) calculates
      (1) List of neighbors at the instant t.
      (2) The amount of remain battery
      (3) Direction of movement of a node
      (4) Speed of movement
      (5) Weight calculation based on WCA algorithm
(4) Compare the weights of normal nodes
(5) Select the node with the minimum weight as Cluster head.

4   Conclusion

Due to the essential role in mobile ad hoc network, selecting the adequate cluster head is a difficult task and important challenges. Clustering is one of the techniques that are used to organize more scalable networks. Although, providing security issues in clustering algorithms had been neglected in the literature, recently trust and reputation based solutions have been proposed to protect clustering algorithms against insider attacks and malicious behaviors of compromised nodes. In this paper, we reviewed the most prominent algorithms proposed in the literature on the basis of trustiness and WCA in manets. The main purpose of every algorithm is to select the adequate node by eliminating the malicious node from clustering process and choosing the node with the minimum weight to be cluster head.

References

1. Hussein, A., Abu Salem, A.O., Youssef, S.: A flexible weighted clustering algorithm based on battery power for mobile ad hoc networks, pp. 2012–2017 (2008). https://doi.org/10.1109/isie.2008.4677234

2. Preetha, V., Chitra, K.: Clustering and cluster head selection technique in mobile ad-hoc network. Int. J. Innov. Res. Comput. Commun. Eng. **2**(7), 5151–5157 (2014)
3. Ait Rahou, M., Hasbi, A.: Toward cluster head selection criterions in mobile ad hoc network. In: Noreddine, G., Kacprzyk, J. (eds.) International Conference on Information Technology and Communication Systems, ITCS 2017. Advances in Intelligent Systems and Computing, vol. 640 (2018)
4. Kwon, T.J., et al.: Efficient flooding with passive clustering an overhead-free selective forward mechanism for ad hoc/sensor networks. Proc. IEEE **91**(8), 1210–1220 (2003)
5. Chatterjee, M., Das, S.K., Turgut, D.: WCA: a weighted clustering algorithm for mobile adhoc networks. Clust. Comput. J. **5**(2), 193–204 (2002)
6. Maleknasab, M., Bidaki, M., Harounabadi, A.: Trust-based clustering in mobile ad hoc networks: challenges and issues. Int. J. Secur. Appl. **7**, 321–342 (2013). https://doi.org/10.14257/ijsia.2013.7.5.30
7. Yu, Y., Zhang, L.: A secure clustering algorithm in mobile ad hoc networks. In: 2012 IACSIT Hong Kong Conferences, vol. 29 (2012)
8. Dahane, A., Berrached, N., Kechar, B.: Energy efficient and safe weighted clustering algorithm for mobile wireless sensor networks. Procedia Comput. Sci. **34**, 63–70 (2014). https://doi.org/10.1016/j.procs.2014.07.040. ISSN 1877-0509
9. Gupta, N., Singh, R.K., Shrivastava, M.: Cluster formation through improved weighted clustering algorithm IWCA for mobile ad-hoc networks. In: 2013 Tenth International Conference on Wireless and Optical Communications Networks (WOCN), Bhopal, pp. 1–5 (2013)
10. Ashwin, M., Kamalraj, S., Azath, M.: Wirel. Personal Commun. **94**, 2203 (2017). https://doi.org/10.1007/s11277-016-3371-0
11. Elhdhili, M.E., Ben Azzouz, L., Kamoun, F.: CASAN: clustering algorithm for security in ad hoc networks. Comput. Commun. **31**(13), 2972–2980 (2008). https://doi.org/10.1016/j.comcom.2008.04.001. ISSN 0140-3664. http://www.sciencedirect.com/science/article/pii/S0140366408002077 (2018)
12. Gomathi, K., Parvathavarthini, B.: A secure clustering in MANET through direct trust evaluation technique. In: International Conference on Cloud Computing (ICCC), Riyadh (2015). https://doi.org/10.1109/CLOUDCOMP.2015.7149624
13. Peng, S., Jia, W., Wang, G.: Voting-based clustering algorithm with subjective trust and stability in mobile ad hoc networks. In: IEEE/IFIP International Conference on Embedded and Ubiquitous Computing, Shanghai, pp. 3–9 (2008). https://doi.org/10.1109/euc.2008.93
14. Kadri, B., Mhamed, A., Feham, M.: Secured clustering algorithm for mobile ad hoc networks. Int. J. Comput. Sci. Netw. Secur. **7**(3), 27–34 (2007)

The Benefits of SDN Integration on 5G Mobile Network

Souad Faid[✉] and Mohamed Moughit

Laboratory of Computer Networks, Mobility and Modeling IR2M,
Settat, Morocco
faidsouad28@gmail.com

Abstract. The current growth in wireless network proof that there is no signs of slowing down. The 5G technology is in its first steps research, in this way researches are currently in progress to explore different architectural designs to address their key challenges. SDN technology have been seen as promising enablers for this new motivation of networks development, which will play an essential role in the design of 5G wireless networks. Existing wireless networks are starting to be insufficient in meeting the actual data demand, for the reason of their inflexible and expensive equipment as well as complex and non-agile control plane management. The gain of Software-Defined Networking (SDN) technology has increased in recent years, in the process to change net-working ecosystems. It has already been applied to data center networks and wide area networks. This paper found that the software-defined approach is thus a reasonable candidate architecture to offer, in part data demand as well as the quality of service for these data. We considered that software defined networks (SDNs) is the best solution. As the simulation evaluation results suggest, the proposed architecture can effectively improve the quality of service in terms of: time insertion, Latency and throughput.

Keywords: SDN · 5G technology · QOS

1 Introduction

In recent times we have hug disruptive trends technologies advances in mobile communications in terms of the number of terminal devices and services offered over a wireless network. Moreover, the increase in the number of mobile devices will be varied exponentially with the explosion of the Internet of Things (loT), including, among others, wearable computers and Machine to Machine (M2M) communications. These offer content and applications that could not be supported by previous generations of mobile devices, such as VoIP, video conferencing and IPTV It is expected that monthly global mobile data traffic will be more than 15 exabytes by 2018 [1], and future 5G mobile networks will need to be able to support this volume of traffic with the needed QOS. We promote that a key differentiator of 5G systems from 4G will be in how we design the overall system control to realize the benefits of virtualization while taking the full advantage of the transport capacity distributed over a large geographical surface [2].

© Springer Nature Switzerland AG 2020
M. Ezziyyani (Ed.): AI2SD 2019, LNNS 92, pp. 266–276, 2020.
https://doi.org/10.1007/978-3-030-33103-0_27

2 State of the Art 5G Technology

2.1   5G Technology Challenges

5G technology is an end-to-end ecosystem to enable a fully mobile and connected society. It empowers value creation towards customers and partners, through existing and emerging use cases, delivered with consistent experience, and enabled by sustainable business models.

The three fundamental requirements for building 5G wireless networks are:

- Capabilities for supporting massive capacity and massive connectivity;
- Support for an increasingly diverse set of services, application and users, all with extremely diverging requirements for work and life;
- Flexible and efficient use of all available non-contiguous spectrum for wildly different network deployment scenarios.

Based on the design principles, the 5G architecture is a native SDN/NFV architecture covering aspects ranging from devices (mobile/fixed) infrastructure, network functions, value enabling capabilities and all the management functions to orchestrate the 5G system. APIs are provided on the relevant reference points to support multiple use cases, value creation and business models [3]. This architecture is illustrated in Fig. 1.

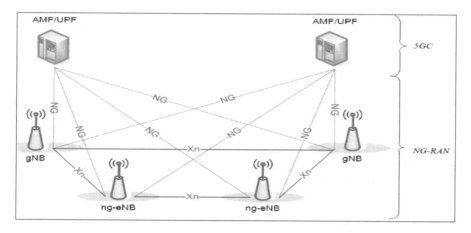

Fig. 1. The architecture of 5G technology

The architecture comprises three layers and an E2E management and orchestration entity:

- The infrastructure resource layer consists of the physical resources of a fixed-mobile converged network. The resources are exposed to higher layers and to the end-to-end management and orchestration entity through relevant APIs.

- The business enablement layer is a library of all functions required within a converged network in the form of modular architecture.
- Specific applications and services of the operator, enterprise, verticals or third parties that the business application layer contains utilize the 5G network (Fig. 2).

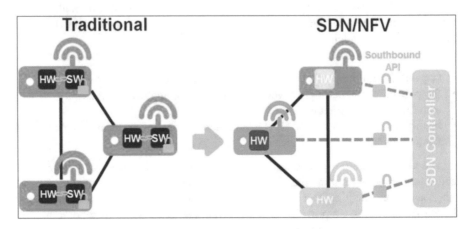

Fig. 2. The decoupling of SDN & NVF

The future challenge of 5G technology is solving the radio capacity problem, in this way the revolution is the software-defined networks and network function virtualization [4].

2.2   Concepts for 5G Mobile Networks

The 5G terminals will have software defined radios and modulation schemes as well as new error-control schemes that can be downloaded from the Internet. The development is seen towards the user terminals as a focus of the 5G mobile networks. The terminals will have access to different wireless technologies at the same time and the terminal should be able to combine different flows from different technologies.

Now, we will go through all OSI layers (Fig. 3) in the 5G mobile terminal design [5].

- Physical/MAC layers

Physical and Medium Access Control layers i.e. OSI layer 1 and OSI layer 2, define the wireless technology. For these two layers the 5G mobile networks is likely to be based on Open Wireless Architecture [6].

- Network layer

The IPv4 (version 4) is worldwide spread and it has several problems such as limited address space and has no real possibility for QoS support per flow. These issues are solved in IPv6, but traded with significantly bigger packet header.

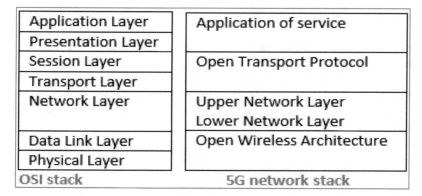

Fig. 3. Protocol stack in 5G technology vs OSI stack

- Open Transport Protocol (OTA) layer

  For 5G mobile terminals will be suitable to have transport layer that is possible to be downloaded and installed. This is called here Open Transport Protocol - OTP.

- Application layer

  Regarding the applications, the ultimate request from the 5G mobile terminal is to provide intelligent QoS management over variety of networks. The QoS parameters, such as delay, jitter, losses, bandwidth, reliability, will be stored in a database in the 5G mobile phone with aim to be used by intelligent algorithms running in the mobile terminal as system processes.

2.3 Quality Requirements in 5G Networks

The highest requirements for experienced user throughput are formed for "Virtual reality office" use case. End-users should be able to experience data rates of at least 1 Gb/s in 95% of office locations and at 99% of the busy period. Additionally, end-users should be able to experience data rates of at least 5 Gb/s in 20% of the office locations. The highest requirements for network latency are formed for "Dense urban information society" use case, device-to-device (D2D) latency is less than 1 ms. the highest requirements for availability and reliability of 5G network are identified for "Traffic safety and efficiency" use case, 100% availability with transmission reliability [6] (Table 1).

Table 1. QOE requirements for 5G networks

QOE indicators	Requirements
Experienced user throughput	5 Gbps in data level (DL) and user level (UL)
Latency	Less than 1 ms
Availability	99%
Reliability	99,9%

3 Software Defined Networks

Introduction of intelligence towards 5G can address the complexity of Heterogeneous Networks by specifying and providing flexible solutions to satisfy for network heterogeneity. Software Defined Networking (SDN) has emerged as a new intelligent architecture for network programmability. The primary idea behind SDN is to move the control plane outside the switches and enable external control of data through a logical software entity called controller.

SDN provides simple abstractions to describe the components, the functions they provide, and the protocol to manage the forwarding plane from a remote controller via a secure channel. This abstraction captures the common requirements of forwarding tables for a majority of switches and their flow tables. This centralized up-to-date view makes the controller suitable to perform network management functions while allowing easy modification of the network behavior through the centralized control plane [7].

3.1 Advantages of SDN in Mobile Network

The main advantages of SDN mobile are:

- Flexible support of middleboxes.
- Better inter-cell interference management.
- Scalable distributed enforcement of QoSand firewall policies in data plane.
- Flexible support of virtual operators by partitioning flow space.
- Seamless subscriber mobility across technologies

In the other hand, the real importance of SDN in mobile cellular is needed in order to:

- Separate concerns.
- Make problems modular and tractable.
- Enable innovative new businesses.

In order to increase the capacity of 5G systems, SDN can provide solutions to overcome the limitations of multi-hop wireless networks [8]. Indeed, SDN can provide advanced caching techniques to store data at the edge network to reach the required high capacity of 5G systems. One way to increase per user capacity is to make cells small and bring the base station closer to the mobile client. In cellular communications, an architecture based on SDN techniques may give operators greater freedom to balance operational parameters, such as network resilience, service performance and QoE. OpenFlow may work across different technologies to provide rapid response to the subscriber mobility and avoid disruptions in the service. The decoupling between the radio network controller and the forwarding plane will enhance the performance of the base station [8] (Fig. 4).

3.2 SDN-Based Design

The SDN controller implements multiple applications such as mobility management, allocation of radio resources and many others, while also interworking with other

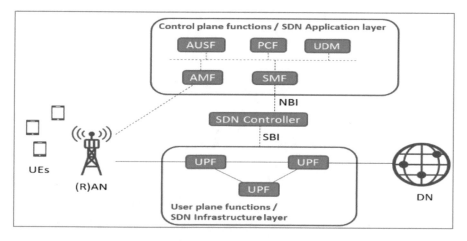

Fig. 4. SDN paradigm over 5G core network

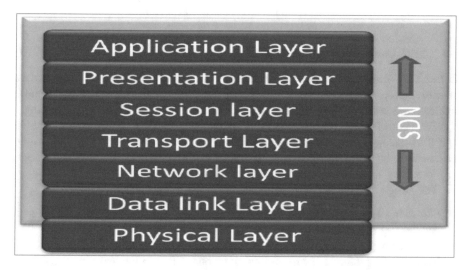

Fig. 5. SDN supports MAC layer to the application layer

programmable elements in the protocol stack. SDN, in the seven layers of OSI, can step over the MAC layer to the application layer, as shown in Fig. 5 [9].

In simple terms, the major contribution of SDN is that the network can be reconstructed. The ideal situation is to achieve a fully automated administration without the design and adjustments of the administrator intervention policy. Policy is the rule for reconstruction, and SDN has a cooperative concept, which indicates easily controlled network problems by defining a good policy set, with a self-learning policy change mechanism. The simple SDN architecture diagram is as shown in Fig. 6 [10].

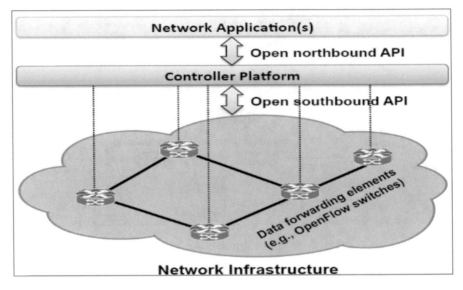

Fig. 6. SDN protocol stack architecture

3.3   SDN Controller Implementation

In the SDN-based design, the control plane is decoupled from the data plane and the control aspects are incorporated in a distributed SDN controller. The dataplane is reduced to a set of SDN-enabled routers for the primary task of packet forwarding.

The control plane, in this decoupled architecture, handles all necessary signaling and also manages the mobility of the UE. We envision the SDN controller implementation to be based on OpenFlow. The SDN controller will push rules to all devices in a proactive manner to handle the large scale of subscribers.

The design of control plane can be summarized in the following steps:

- Separating the control plane functionalities in all network entities starting from eNodeB up to the P-GW.
- Implementing the control plane functions as SDN controller applications.
- Encapsulating the management plane functions within individual virtual machines (VM) that interact with the SDN controller.
- The SDN control plane works in a distributed manner, sharing state among individual controller instances (as can be seen in Fig. 7).

3.4   Integration of SDN for 5G Technology

In a 5G network, the relationship between spectrum and flow will become increasingly obvious as spectrum reuse can relieve traffic between different frequency bands. In other words, if the flow is well-processed, the spectrum will avoid competition, and overloading of bandwidth can also be avoided. Therefore, it is expected that SDN can consider monitoring results when perceiving the frequency spectrum or switching bands.

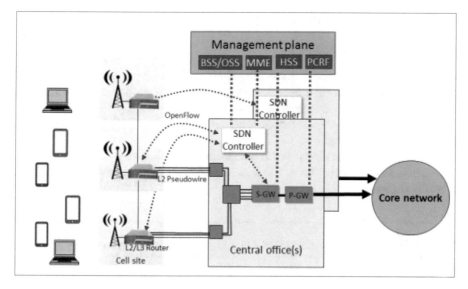

Fig. 7. SDN controller located within the mobile backhaul and its communications with other entities in the network including eNodeB, S-GW and P-GW.

To realize the expected results, this study suggests that the controller in the original SDN architecture should cross the 5G architecture [11].

The following platform essentials are involved in state dissemination, L3 routing, tunnel setup, and interfacing with management plane. The value-added services handle intelligence for automatically managing resources, mobility, DiffServ/QoS and network selection for UE (Fig. 8).

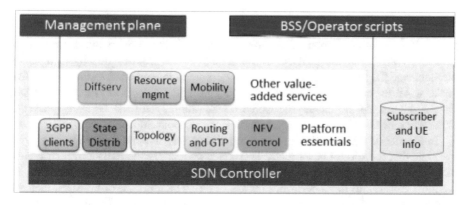

Fig. 8. Design of the SDN controller for the 5G network

4 Benchmark Evaluation Report

4.1 Insertion Time Evaluation

It is undeniable that SDN have the ability to improve network performance. Combining their advantages can create a better architecture, thus, the proposed architecture considers the individual relationships and mutual impacts.

In the following figure you can QOS improvement in term of time insertion with & without SDN implementation.

The Fig. 9 evaluates the proposed architecture and other three conditions in terms of insertion time and delay time. The insertion time represents the time of a service to enter the network.

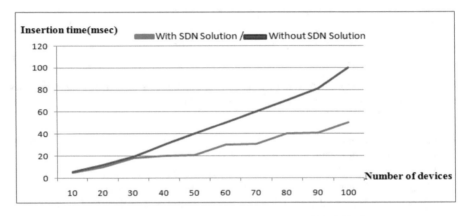

Fig. 9. Relationship between spectrum and bandwidth utilization rate.

4.2 Latency Evaluation

Figure 10 illustrates the total delay time comparison between 5G network using SDN technology a without using SDN. On one hand, we see from Figure, by using SDN mechanisms, the first packet takes 20 ms, i.e. more time compared to the consecutive packets. All the consecutive packets take very less time compared to the first packet. On the other hand the latency measured on Traditional network for every packet is either same or more, the first packet, the latency is about 30 ms.

In 5G mobile network, the spectra computing is very important in order to improve the issue that the spectra is shortage. Therefore, we analyze the impact of interference between whether our proposed mechanism. Of course, we cannot be too intuitive. So we must consider to the factor from network layer which is control by SDN, then we choose the generation rate as our continuous variables. In Fig. 11, we can see the throughput of our proposed mechanism can directly reflect the increasing generation rate when interference is short.

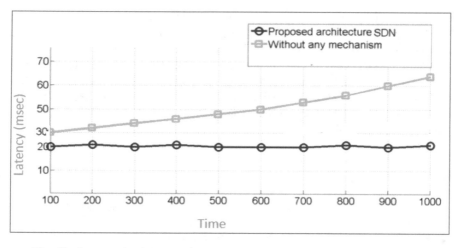

Fig. 10. Latency simulation on SDN using architecture & over traditional network

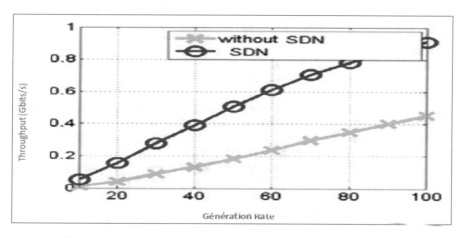

Fig. 11. Relationship between throughput and bandwidth utilization rate

5   Conclusion

The course of digital mobile communication was design to meet the requirements of 5G technology. Basic theoretical knowledge and technologies are illustrated on class.

Because the 5G wireless network is an actual environment, mobile devices are expected to be several times that of the present, thus, there will be inevitable challenges in accessing a network. Meanwhile, the frequency spectrum, bandwidth, security, and various factors pose a trade-off issue.

SDN is capable of creating a programmable network by taking both next-generation systems and existing infrastructure and making them substantially more dynamic. It does this by bringing disparate systems and technologies together under a common management system that can utilize them to their full potential. By using abstraction, SDN can simplify the software needed to deliver services, improving the use of the network and shortening delivery times, leading to increased revenue.

We argue that the new design based on the centralized SDN controller not only allows for programmability and evolution of the mobile networks, but also in the long run reduces various maintenance and upgrade costs of the network. Moreover, the SDN based controller can facilitate several value-added services, including policy based services, for the mobile operators.

References

1. Pandey, M.S., Kumar, M., Panwar, A., Singh, I.: A survey: wireless mobile technology generations with 5G. Int. J. Eng. Res. Technol. **2**(4) (2013)
2. Wang, L.-C., Rangapillai, S.: A survey on green 5G cellular networks. In: Proceedings of the Signal Processing Communication (SPCOM), pp. 1–5, July 2012
3. Gohil, A., Modi, H., Patel, S.K.: 5G technology of mobile communication: a survey. In: Proceedings of the International Conference on Intelligent Systems and Signal Processing (ISSP), pp. 288–292, March 2013
4. Tuttlebee, W.H.W. (ed.): Software Defined Radio: Enabling Technologies. Wiley, New York (2003)
5. Akyildiz, I.F., Lee, W.-Y., Vuran, M.C., Mohanty, S.: NeXt generation/dynamic spectrum access/cognitive radio wireless networks: a survey. Comput. Netw. **50**(13), 2127–2159 (2006)
6. McKeown, N., et al.: OpenFlow: enabling innovation in campus networks. ACM SIGCOMM Comput. Commun. Rev. **38**(2), 69–74 (2008)
7. Lu, H.-C., Liao, W., Lin, F.Y.-S.: Relay station placement strategy in IEEE 802.16j WiMAX networks. IEEE Trans. Commun. **59**(1), 151–158 (2011)
8. Rappaport, T.S., et al.: Millimeter wave mobile communications for 5G cellular: it will work! IEEE Access **1**, 335–349 (2013)
9. Rost, P., et al.: Cloud technologies for flexible 5G radio access networks. IEEE Commun. Mag. **52**(5), 68–76 (2014)
10. Chin, W.H., Fan, Z., Haines, R.: Emerging technologies and research challenges for 5G wireless networks. IEEE Wirel. Commun. **21**(2), 106–112 (2014)
11. Demestichas, P., et al.: 5G on the horizon: key challenges for the radio access network. IEEE Veh. Technol. Mag. **8**(3), 47–53 (2013)

The Impact of Mobility Speed over Varying Radio Propagation Models Using Routing Protocol in MANET

Mahmood Khan[1], Muhammad Faran Majeed[1], Amjad Mehmood[2],
Khalid Saeed[1], and Jaime Lloret[3(✉)]

[1] Department of Computer Science, Shaheed Benazir Bhutto University Sheringal,
Sheringal, Pakistan
{mahmood,khalidsaeed}@sbbu.edu.pk, m.faran.majeed@ieee.org
[2] Institute of Computing, Kohat University of Science and Technology,
Kohat, KP, Pakistan
dramjad.mehmood@ieee.org
[3] Integrated Management Coastal Research Institute,
Polytechnic University of Valencia, Camino Vera s/n, 46022 Valencia, Spain
jlloret@dcom.upv.es

Abstract. Mobile Ad hoc Network (MANET) is temporary network shaped for a particular function such as transferring data from one node to another that can change locations dynamically without any network architecture. This paper primarily focuses on three propagation models used in mobile ad hoc network. These propagation models are examined and evaluated through changing mobility and traffic parameters. The results of varying radio models are analyzed over DSDV (Destination-Sequenced Distance-Vector) routing protocol. The simulation is carried out using Network Simulator-2 in terms of average network delay, delivery ratio, throughput, and packet drop ratio over mobility parameter speed. The results show that these three models perform differently by using same traffic and mobility factors. The research reveals that the effect of varying mobility speed has more impact on the throughput of all the models especially Shadowing and Two Ray models. However, Shadowing model shows better data sending ratio at higher mobility speed than other models. Shadowing model tends to show longer and consistent average delay and drop more data packets than other models at high mobility speed.

Keywords: MANET · Propagation models · DSDV · Random
WayPoint · NS-2 · Mobility speed

1 Introduction

Mobile Ad hoc network is a type of network that can change locations dynamically without any network infrastructure connecting different network nodes.

© Springer Nature Switzerland AG 2020
M. Ezziyyani (Ed.): AI2SD 2019, LNNS 92, pp. 277–288, 2020.
https://doi.org/10.1007/978-3-030-33103-0_28

Wireless ad hoc networks are made up of independent nodes that are self-managed without any network system. Thus, nodes in ad hoc networks can easily join or leave the network due to dynamic nature of its topology [1]. A number of protocols have been suggested for such type of networks e.g. Dynamic Source Routing (DSR), Ad hoc On-demand Distance Vector (AODV) and Temporally Ordered Routing Algorithm (TORA) etc [2].

There are three most important types of ad hoc routing protocols, that are proactive, reactive and hybrid. In proactive AKA table-driven routing protocols, the nodes establish the routing paths from the beginning rather than on demand and remain stable for the longer time until there is change in topology. In reactive routing protocols, the routes are managed temporarily, are established on demand and maintained until new routes are discovered [14]. Reactive routing protocols use low power, have low bandwidth consumption, and less resources requirements at the time of communication. Hybrid routing protocols have combined characteristics of both proactive and reactive routing protocols. The hybrid routing protocols demonstrate better performance over a large network of mobile nodes in cluster or grid form.

1.1   Destination-Sequenced Distance-Vector Routing (DSDV)

DSDV routing protocol is a proactive ad hoc routing protocol in which routing tables are updated whenever change occurs. This change is in the form of a node joining or leaving network, availability of path, and accessibility of network [3]. Routes are maintained with a sequence number and new routes are replaced with old routes acquiring recent sequence numbers.

The sequence numbers are regenerated by the destination and are sent to the source for the transmission of the packets. The routing information is sent using full dumps and incremental dumps [13].

1.2   Radio Propagation Models (RPM)

RPM are based on phenomenon in which radio waves are propagated from one place to another. Different kinds of radio propagation models are used in the field of wireless ad hoc networks. The most popular propagation models that are used in the research work are Two Ray Ground model, Free Space model and Shadowing model.

Two Ray Ground model is frequently used radio propagation model. This model performs better and gives more accurate results when long distance transmissions are covered [4]. Free Space model (FSM) is used when there is a short distance and line of sight path between transmitter and receiver. Shadowing model made of two components, the first part assumes the mean received power at distance while the second part reflects the change of received power at given distance [22].

1.3   Objective

To analyze the impact of these three propagation models along with Destination-Sequenced Distance-Vector routing protocol by mobility parameter (speed) using various metrics.

2   Related Work

Mobile ad hoc networks and routing protocols have been examined and studied broadly by different researchers in the recent past and have added great information to a large extent in the existing mine of comprehension as well as found some new openings in this area. Some of the appreciable researches in this regard are given below.

Khider et al. [4] analyzed the impact of different radio propagation models in urban area using NS-2 for MANET. The work was based on comparing different metrics such as packet delivery ratio, packets dropped, throughput, packets sent, and packet control overhead. Using MATLAB tool, the AODV routing protocol was analyzed using propagation models under Random WayPoint (RWP) protocol. The performance was analyzed by varying mobility speed. The results exhibit that Free Space and Two Ray models show high throughput, packets sending ratio, and packet delivery ratio as compared to Shadowing model. They observed that with increase in the mobility speed, the control overhead using Two Ray model had more impact on the performance of Shadowing and Free Space models.

Usop et al. [5] presented a Mobile Grid Model that is newly established in wireless network and properly shifted from conventional network. They compared the performance analysis of different ad hoc protocols. The reactive routing protocol such as Ad hoc On-Demand Distance Vector (AODV) performed well using Transmission Control Protocol (TCP) based traffic. However, AODV node delivers small amount of packets at high mobility speed and failing to converge according to the network. On the other hand, Dynamic Source Routing (DSR) protocol performed well, considering all mobility rates. Contrary, DSDV also performed fairly equal to that of DSR but have a disadvantage of high routing or control overhead of packet transmissions. Also DSDV behaved poorly at higher mobility rates than DSR.

Eenennaam [6] surveyed distinct propagation models used in VANET based on simulation studies. The author presented and raised several key issues based on propagation model used in such type of networks and noticed large influence on the propagation models. For Example, nodes which are able to communicate and have right reception probability are affected most. He also noticed the increase in delay in a multi-hop scenario due to speed and also packets' propagation throughout the network. It was also shown that the overhead with respect to collisions and medium utilization is also affected by the probability distribution of correct reception. The assessment exhibit that the implementation of the real world scenario could act in a diverse way from the simulation, thus the model and parameters should be properly mapped to the target environment.

For example, deterministic site-specific propagation modeling and site planning of cellular systems is not appropriate for VANET simulations' nature of dynamic environment and the change in the mobility, because the modeling of propagation environment is done in stochastic way.

Shah et al. [7] attempted to compare the performance of two popular frequently used on-demand or reactive routing protocols for MANETs. The simulation was performed by taking MAC and physical layer models to study the layer-to-layer interactions and also to identify their inferences. The results showed that AODV and DSR outperformed DSDV protocol. However, there is insignificant dissimilarity between the working mechanism of DSR and AODV, contribute few similarities of on-demand nature but minor differences between these protocols' mechanisms can lead to significant distinctions in their performance. The evaluation was performed by taking variety of traffic workload and mobility scenarios in terms of speed, source traffic load, and network size. The performance discrepancies were analyzed over varying quantities of above mentioned parameters and results were determined using NS-2 simulator.

Divecha et al. [8] evaluated the impact of various mobility models on the working mechanism of two ad hoc routing protocols DSR and DSDV. These protocols' effect was studied and analyzed under four mobility models by considering different scenarios. These were RWP, Group Mobility, Freeway and Manhattan models. The performance of these protocols is evaluated by changing the node and hops between the source and destination in specific area. The experimental results verified that the performance of these protocols depends upon varying mobility models, node densities, and number of hops. Thus, they have selected and considered the specific mobility models suiting a specific application by choosing a peculiar routing protocol. The results showed that DSR gives improved results in a highly dynamic and mobile network than DSDV, while DSR route convergence and quest for finding and determining the new routes up to the destination is notable.

Muktadir [9] investigated the MANET routing protocols under energy utilization for proactive protocol named OLSR and reactive protocol named Dynamic MANET On-demand Routing Protocol (DYMO). The effect of Manhattan Grid Mobility Model is simulated and analyzed using NS-2. Using this mobility model, the nodes moved according to a specified pattern in a dense scenario in urban area. The energy consumption by the nodes was calculated mathematically. The results were calculated in terms of the energy consumption and throughput under the two routing protocols by changing node speed, transmission range, and packet sending rate. It has been observed that the Optimized Link State Routing Protocol (OLSR) consumes less energy and found to be more capable than DYMO. The change in speed has great effect on the performance of DYMO than OLSR.

Gauthier et al. [10] compared the different routing protocols on the basis of radio propagation model. The simulation results exposed that OLSR is more efficient than AODV in terms of throughput. However, AODV also performed well as a reactive routing protocol when the mobility speed increases. While in case of routing load, OLSR showed less control load than AODV. They studied the effect of the external interferences on working of the ad hoc network and investigated four ad hoc routing protocols. They also addressed the results over interference levels and MAC frames. Several conclusions were drawn in this research work in building cross-layer algorithms by considering many physical parameters. They observed that the quantity of interferences is directly affecting the quality of the route and also the optimality of hops and traffic parameters, and so on. The results finally pictured that OLSR presented improved throughput and delay than AODV. But in case of high mobility, OLSR showed a smaller amount interference and low control load. Thus it was evaluated that OLSR performs better than AODV even at increased speed and throughput. AODV also performed better at high mobility speed.

Gruber et al. [11] adapted the propagation model named Walfisch-Ikegami that is used commonly for doing simulations in urban area scenarios. The most exceptional feature of this propagation model is that it permits acceptable propagation predictions but it does not help in delay computation. They noticed that ad hoc protocols are capable to work in this environment, but in case of simulation through flat environments, the network performance is degraded as a whole. They used an enhanced and sophisticated version of the random waypoint mobility model to allow reasonable simulation environment. They observed that the AODV and DSR performance remains unaffected. The results illustrated that DSR behaved excellently than AODV, when the network is small and mobility speed is low. On the contrary, AODV performed well in scenarios with large number of nodes and high mobility speed.

3   Research Methodology

This paper reveals MANETs in the light of various propagation models and the analysis of myriad routing protocols in order to enhance network performance. On the basis of diverse literature reviews and researches, different traffic and mobility types were examined over distinct propagation models under ad hoc routing protocols. The NS-2 has been used for conducting the simulations. These propagation models are tested by keeping *speed* as main parameter in scenarios. The results were extracted from the simulations using performance metrics such as packet success ratio, number of packets sent/dropped, throughput and average network delay. After running simulations, the results were obtained which were plotted into graphs using MATLAB.

3.1   Random WayPoint Mobility Model (RWP)

Random WayPoing Mobility Model (RWP) is entity mobility model and considered by unpredicted movement and rapid discontinuation. It includes mobility speed and change in direction. In RWP, the mobile nodes choose a random target and moves towards it with a speed which is dispersed uniformly between maximum and minimum speed. Upon reaching the destination, a node pause for instant which is defined by a constant or uniform pause time (for example in seconds) [12, 15, 16], and [17] and again selects and move towards a new destination and the process continues until it reaches the ultimate destination. RWP forms a zigzag style mobility pattern and the nodes converge, disperse and converge again till the end of the simulation.

3.2 Tools Used

There is a range of network simulation tools available having distinctive features in order to simulate the diagnosed problem. Examples of such tools are OPNET, MATLAB, Qual net, GloMoSim and NS-2. In this research work, NS-2 has been used to examine the performance of the three propagation models over DSDV routing protocol.

NS-2 is a big library of network and protocol objects. Its applications include such as (web and File Transfer Protocol (FTP) etc.), protocols (transport and routing), network media types (satellite links, wired and wireless) and network elements (mobile nodes and wireless channel models). NS-2 is composed of object oriented methodology and Extension of Object Tool Command Language TCL (OTcl) interpreter. For higher efficiency in simulation these objects are written in C++.

For viewing the simulation results and packet tracing we used Network Animator is TCL based tool. It cover the simulation of topology layout, packet level animation.

AWK is a scripting type programming language based on C/C++. AWK language has used for manipulation and processing of text based data. MATLAB is used for mathematical calculations, creating different types of graphs and making pivot tables used in this paper.

4   Experimental Results and Analysis

The three radio propagation models such as Two Ray Ground, Free Space and Shadowing Models are analyzed on the basis of different traffic and mobility factors in the light of following performance metrics such as packet success ratio, throughput, packets sent, packets dropped ratios, and average network delay [18] and [19] (Table 1).

4.1   Scheme-1

Scheme-1 explains the impact of mobility speed using different metrics such as packets success ratio, throughput, packets sent, packets dropped ratio, and average network delay.

Figure 1 shows the *Packets Success Ratio* with respect to variation in mobility speed for 50 nodes using Two Ray, Free Space and Shadowing Propagation Models for same area. The graph shows that the effect of mobility speed changes on performance of wireless models when the pause time is 400 s. The packet

Table 1. Packets success ratio vs. mobility speed

Speed (pause time 400 s)	Two ray ground	Free space	Shadowing
5	98.03	98.37	99.26
10	96.48	95.35	95.31
20	97.53	96.06	96.09
25	95.93	97.66	98.38

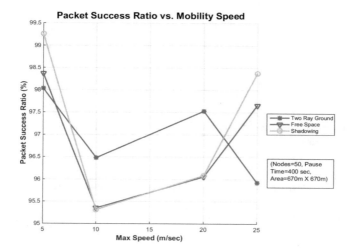

Fig. 1. Packets success ratio vs. mobility speed for two ray ground, free space and shadowing propagation models.

delivery fractions for Free Space and Shadowing models are almost stable with enlarging the speeds, while it has a small changes for Two Ray model. Shadowing model shows higher packet success ratio than Free Space and Two Ray models by about 97.5% and 96% at 25 m/s respectively (Table 2).

Figure 2 shows the *Throughput* with respect to mobility speed for 50 nodes using Two Ray, Free Space and Shadowing Propagation models. The effect of varying mobility speed has more impact on the throughput of all the models especially Shadowing and Two Ray when pause time is fixed to 400 s. However,

Table 2. Throughput vs. mobility speed

Speed (pause time 400 s)	Two ray ground	Free space	Shadowing
5	1996	2037	2026
10	1976	1973	1929
20	2016	1968	2060
25	1957	2006	2055

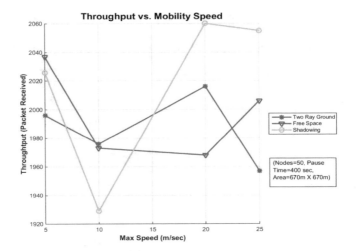

Fig. 2. Throughput vs. mobility speed for two ray ground, free space and shadowing propagation models.

Table 3. Packets send (data) vs. mobility speed

Speed (pause time 400 s)	Two ray ground	Free space	Shadowing
5	2036	2052	2026
10	2048	2070	2023
20	2067	2048	2060
25	2040	2039	2055

Shadowing shows highest throughput than the other models at mobility speed of 25 m/s (Table 3).

Figure 3 shows *Number of Data Packets Sent* with respect to mobility speed for 50 nodes using different models. The graph shows that Free Space and Two Ray models tend to send more data packets from CBR sources to destinations at mobility speed of 10 and 20 m/s at constant pause time of 400 s. However, Shadowing shows good data sending ratio at higher mobility speed than the other models.

Figure 4 shows the *Number of Data Packets Dropped* with respect to mobility speed using different models at constant pause time of 400 s. The graph shows that Shadowing tends to drop more data packets than Free Space and Two Ray models especially at mobility speed of 10 m/s (Tables 4 and 5).

Figure 5 shows the *Average Network Delay* with respect to mobility speed for different models. The graph shows that the Shadowing tends to show longer and consistent average delay throughout the mobility speed than the other models. On the other hand, Two Ray Ground model shows notable delay between mobility speed of 10 and 20 m/s respectively.

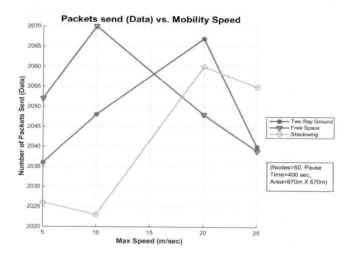

Fig. 3. Number of data packets sent vs mobility speed for two ray ground, free space and shadowing propagation models.

Table 4. Packets dropped (data) vs. mobility speed

Speed (pause time 400 s)	Two ray ground	Free space	Shadowing
5	83	30	224
10	146	194	383
20	103	159	329
25	169	66	266

Table 5. Average network delay vs. mobility speed

Speed (pause time 400 s)	Two ray ground	Free space	Shadowing
5	0.046	0.051	0.063
10	0.041	0.055	0.058
20	0.064	0.054	0.059
25	0.039	0.054	0.057

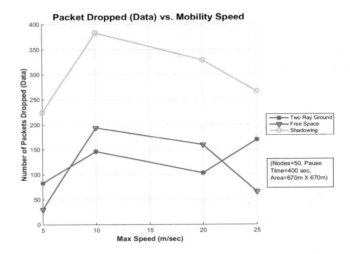

Fig. 4. Number of data packets dropped vs. mobility speed for two ray ground, free space and shadowing propagation models.

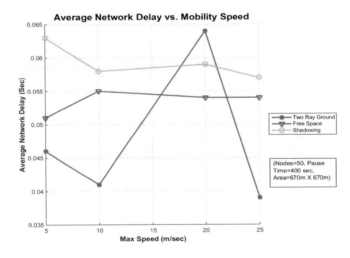

Fig. 5. Average network delay vs. pause time for two ray ground, free space and shadowing propagation models.

5   Conclusion and Future Work

The research unmasks the results of the three mobile ad hoc models (Two Way Ground, Free Space and Shadowing Models) on DSDV Routing Protocol using Random WayPoint Mobility Model. The outcomes show that the three models performed differently by using same traffic and mobility speed. The change in speed has extra effect on the performance of packet success ratio on all the models.

In case of throughput, Two Ray and Free Space showed good success ratio in terms of number of packets received and, in some cases, Shadowing model also showed good value of throughput at high mobility speed. Free Space model showed good numbers while Shadowing model exhibited drastic increase in number of data packets dropped as compared to the others at high mobility speed. In case of average network delay, Two Ray Ground model has shown fluctuations in delay value at high mobility speed as compared to others while Shadowing model has shown longer and consistent delay at high mobility speed. The key conclusion of the research is that Two Ray Ground and Free Space Propagation Models suit DSDV protocol with Random WayPoint Model considering urban area, with buildings, trees and hills etc. However, Free Space Propagation model suits roof top, building top and above the ground area with clear line of sight between transmitter and receiver. On the other hand, Shadowing model has good packets success ratio and throughput but has drawback of high packet drop and average network delay. In our future work, we are going to include in our model indoor environments [20, 21].

References

1. Rattan, K., Patel, R.B.: Mobile ad hoc networks: challenges and future. In: Proceedings of National Conference on Challenges and Opportunities in Information Technology, RIMT, Gobindgarh, Punjab, India, pp. 133–135 (2007)
2. Park, V.D., Corson, M.S.: A highly adaptive distributed routing algorithms for mobile wireless networks. In: Proceedings of the IEEE INFOCOM, vol. 3, pp. 1405–1413 (1997)
3. Perkins, C.E., Bhagwat, P.: Highly Dynamic Destination-Sequenced Distance-Vector Routing (DSDV) for Mobile Computers, pp. 234–244. ACM (1994)
4. Khider, I., Wang, F., Yin, W., Sacko: The impact of different radio propagation models for mobile ad-hoc networks (MANET) in urban area environment. World Journal of Modeling and Simulation ISSN 1 746-7233, England, UK. 5, pp. 45–52 (2009)
5. Usop, N.S.M, Abdullah, A., Abidin, A.F.A.: Performance evaluation of AODV, DSDV & DSR routing protocol in grid environment. Int. J. Comput. Sci. Netw. Secur. (IJCSNS) **9**(7), 261–268 (2009)
6. Eenennaam-van, M.E.: A survey of propagation models used in vehicular ad hoc network (VANET) research. Design and Analysis of Communication Systems group, Faculty of EEMCS, University of Twente, Netherlands (2008)
7. Shah, S., Khandre, A., Shirole, M., Bhole, G.: Performance evaluation of ad hoc routing protocols using NS2 simulation. Veermata Jijabai Technological Institute, Mumbai, India. Mobile and Pervasive Computing (CoMPC) (2008)
8. Divecha, B., Abraham, A., Grosan, C., Sanyal, S.: Analysis of dynamic source routing and destination-sequenced distance-vector protocols for different mobility models. In: Proceedings of the First Asia International Conference on Modeling & Simulation (AMS 2007), pp. 224–229. IEEE 0-7695-2845-7/07 (2007)
9. Al-Muktadir, H.A.: Energy consumption study of OLSR and DYMO MANET routing protocols in urban area. In: National Conference on Communication and Information Security (NCCIS). Daffodil International University, Dhaka, Bangladesh (2007)

10. Gauthier, V., De-Rasse, R., Marot, M., Becker, M.: On a comparison of four ad-hoc routing protocols when taking into account the radio interferences. In: 3rd International Working Conference on Performance Modeling and Evaluation of Heterogeneous Networks (HET'NETs 2005), England (2005)
11. Gruber, I., Knauf, O., Li, H.: Performance of ad hoc routing protocols in urban environments. Lehrstuhl für Kommunikationsnetze Technische Universität München, Siemens AG Information Communication Mobile (2004)
12. Camp, T., Boleng, J., Davies, V.: A survey of mobility models for ad hoc network research. Department of Math and Computer Sciences Colorado School of Mines, Golden, CO, vol. 2, no. 5, pp. 483–502 (2002)
13. Majeed, M.F., Ahmed, S.H., Dailey, M.N.: Enabling push-based critical data forwarding in vehicular named data networks. IEEE Commun. Lett. 21(4), 873–876 (2016)
14. Khan, M., Majeed, M.F., Muhammad, S.: Evaluating radio propagation models using destination-sequenced distance-vector protocol for MANETs. Bahria Univ. J. Inf. Commun. Technol. (BUJICT) 10(1) (2017)
15. Garcia, M., Catalá, A., Lloret, J., Rodrigues, J.J.P.C.: A wireless sensor network for soccer team monitoring. In: International Conference on Distributed Computing in Sensor Systems (DCOSS 2011), Barcelona, Spain, 27–29 June 2011 (2011)
16. Lloret, J., Garcia, M., Catala, A., Rodrigues, J.J.P.C.: A group-based wireless body sensors network using energy harvesting for soccer team monitoring. IJSNet 21(4), 208–225 (2016)
17. Ullah, I., Shah, M.A., Wahid, A., Mehmood, A., Song, H.: ESOT: a new privacy model for preserving location privacy in internet of things. Telecommun. Syst. 67(4), 553–575 (2018)
18. Zhang, Z., Mehmood, A., Shu, L., Huo, Z., Zhang, Y., Mukherjee, M.: A survey on fault diagnosis in wireless sensor networks. IEEE Access 6, 11349–11364 (2018)
19. Mehmood, A., Alrajeh, N., Mukherjee, M., Abdullah, S., Song, H.: A survey on proactive, active and passive fault diagnosis protocols for WSNs: network operation perspective. Sensors 18(6), 1787 (2018)
20. Lloret, J., López, J.J., Turró, C., Flores, S.: A fast design model for indoor radio coverage in the 2.4 GHz wireless LAN. In: 1st IEEE International Symposium on Wireless Communication Systems, pp. 408–412 (2004)
21. Sendra, S., Fernandez, P., Turro, C., Lloret, J.: IEEE 802.11 a/b/g/n indoor coverage and performance comparison. In: 6th International Conference on Wireless and Mobile Communications (ICWMC 2010), Athens, Greece, 28 June 2020–02 July 2020 (2020)
22. Ghafoor, K.Z., Abu Bakar, K., Lloret, J., Khokhar, R.H., Lee, K.C.: Intelligent beaconless geographical forwarding for urban vehicular environments. Wirel. Netw. 19(3), 345–362 (2013)

Thing-Based Service-Oriented Architecture for Industry 4.0

Fatima Zohra El Kho[✉] and Noura Aknin[✉]

TIMS Research Unit, LIROSA Laboratory, Abdelmalek Essaadi University,
Tetouan, Morocco
fatima.zohra.elkho@gmail.com, aknin@uae.ma

Abstract. The evolution of ICT helps the organizations to step forward and generate value. In fact, Internet helps businesses to break down physical and technological limitations and establish global networks that incorporate their machinery, warehousing systems and production facilities in the shape of Cyber-Physical Systems (CPS). In the manufacturing environment, these CPS comprise smart machines, storage systems and production facilities capable of autonomously exchanging information, triggering actions and controlling each other independently thanks to Internet of Thing and Services (IoTS), Cloud Computing and Cognitive Computing capabilities. This paper will present a state of art of this technological evolution that can be described as new industrial revolution, or Industry 4.0. Then we propose a new IoT-based Service Oriented Architecture for smart industry application, focusing on integration challenges imposed by IoT generating extreme heterogeneity and large amount of data to process.

Keywords: Internet of Things (IoT) · Industry 4.0 (I4.0) · Cyber-Physical Systems · Service Oriented Architecture (SOA) · Integration · Middleware

1 Introduction

To survive in a competitive market, companies should improve their business models and processes.

The digitalization of the global economy is no longer a prophecy, it is now not only a proven fact, but especially a major trend of globalization. In the realm of manufacturing, this technological revolution follows tree stages to industrialization, therefore it can be described as the fourth stage, or Industry 4.0.

Industrialization began with the introduction of mechanical manufacturing equipment at the end of the 18th century in 1784, when machines like the mechanical loom revolutionized the way goods were made. This first industrial revolution was followed by a second one that began around the turn of the 20th century and involved electrically-powered mass production of goods based on the division of labor. This was in turn superseded by the third industrial revolution that started during the early 1970s and has continued right up to the present day. This third revolution employed electronics and information technology (IT) to achieve increased automation of manufacturing

© Springer Nature Switzerland AG 2020
M. Ezziyyani (Ed.): AI2SD 2019, LNNS 92, pp. 289–300, 2020.
https://doi.org/10.1007/978-3-030-33103-0_29

processes, as machines took over not only a substantial proportion of the "manual labor" but also some of the "brainwork".

The forth industrial revolution, I4.0 strategy, aims to ensure an industry fit for future manufacturing challenges. It supports the integration of Cyber Physical Systems (CPS) and Internet of Things and Services (IoTS) with an eye to enhance productivity, efficiency and flexibility of production processes and thus economic growth [1, 2].

This paper is structured as follows. In Sect. 2 we discuss ongoing I4.0 projects and future manufacturing vision all over the world. Then in Sect. 3 we shift gears to discuss multiple paradigms related to I4.0 and we propose a novel Thing-Based Service-Oriented Architecture for I4.0. Finally, we propose a thing-based architecture respecting the IoT requirements such as high availability, security and scalability.

2   Ongoing Industry 4.0 Projects

2.1   Worldwide I4.0 Projects

Industry 4.0 was one of the future projects adopted in the Action Plan High-tech strategy 2020" by the German Federal Government in 2010. The main ideas of Industry 4.0 have been firstly published by KAGERMANN in 2011 and have built the foundation for the Industry 4.0 manifesto published in 2013 by the German National Academy of Science and Engineering (Acatech) [1, 3]. At European level, the Public-Private Partnership (PPP) for Factories of the Future (FoF) addresses and develops Industry 4.0-related topics [4]. Then in 2015, Angela Merkel, German chancellor, spoke glowingly of the concept at the World Economic Forum in Davos, calling "Industry 4.0" the way that we "deal quickly with the fusion of the online world and the world of industrial production". To that end, the German government is investing some €200 million (around £146 million, $216 million, or AU$278 million) to encourage research across academia, business and government, and Germany isn't the only country where advancements are taking place. Today, a total of over 300 players from 159 organizations are active in the platform. The platform also maintains numerous links with stakeholders outside Europe. One example is the Standardization Council i4.0 to initiate and coordinate standards for digital production. The platform is also co-operating with other countries [4, 5]:

- USA: Industrial Internet Consortium
- France: Alliance Industrie du Futur
- Japan: Robot Revolution Initiative
- China: Memorandum of understanding (MoU)

While in **Germany**, a consulting body of the Federal government - The Economy-Science Research Union requested the associations to setup the Industry 4.0 platform, big companies in the **USA** triggered the start: In March 2014 AT&T, Cisco, General Electric, IBM and Intel founded the Industrial Internet Consortium (IIC) in order to coordinate the priorities for the industrial Internet, and to enable the technical applications required for this. Meanwhile 250 companies have joined the movement, including some from Germany. The aim of the Industrial Internet Consortium as described by Dr. Richard Mark Soley, Executive Director of IIC, is to bring together "operational systems", which mean machines and industrial plants in the widest sense of the term, and information technology, as stated in the Machine Market sibling magazine Elektrotechnik [6].

On 18 April 2013, the President of the **French** Republic launched the Innovation 2030 Commission, chaired by Anne Lauvergeon and under the aegis of the Minister for the Economy, Industry and the Digital sector and the Minister Delegate with responsibility for Small and Medium-sized Enterprises, Innovation and the Digital Economy [7].

The **UK**'s vote to leave the EU has been a game changer for UK manufacturing. In December 2016 and early January 2017, two thirds of manufacturing executives said that the uncertainty from Brexit would be bad for UK economic stability. The Aerospace Growth Partnership, for example, which was founded in 2010–11, has helped develop the right skills, supply chain capability and technology to help the sector maintain market share. And government and industry funding, to the tune of £4 billion, has been secured till 2026. "There is no other innovation pot in the UK that is set out for that long", says Jeegar Kakkad, Chief Economist, ADS. To encourage the British companies to have more productivity by using automation, the government has adopted an industrial strategy to 2020, to provide the support to all sectors of activities, mainly concerning the Robotics and Autonomous Systems (RAS) [8].

In 2015, **China** published un industrial development plan, which is called "Made in China 2025" [9]. He goal is to comprehensively upgrade Chinese industry, making it more efficient and integrated so that it can occupy the highest parts of global production chains. The plan identifies the goal of raising domestic content of core components and materials to 40% by 2020 and 70% by 2025 [9, 10].

Indian manufacturing enterprises are still in the very early stages of IoT adoption. The recent instances of some automobile manufacturers using sophisticated algorithms to mislead emission control tests, create the need for a wide ranging IOT policy. In addition of other ongoing case study projects in underground mining industry. The aim of this project is to build a more efficient and safer mine workforce and transformation of the underlying processes (e.g. Asset Maintenance, People Safety…) [11].

At the **Japan**-Germany Summit Meeting held in March 2015, Japan and Germany agreed on the advancement of cooperation concerning IoT/Industrie 4.0 in the manufacturing industry. On April 28, 2016, the Ministry of Economy, Trade and Industry (METI) and the Federal Ministry for Economic Affairs and Energy (BMWi), Germany, signed a joint statement regarding the Japan-Germany cooperation on Internet of Things (IoT)/Industrie 4.0: The Robot Revolution Initiative of Japan and the Platform Industrie 4.0, Germany [12].

In 2017, the **Russian** Government approved the first roadmap to develop the National Technology Initiative – Advanced Industrial Technologies (TechNet). In the Global Competitiveness Index 2017 by the WEF, Russia has risen to 43rd place due to high quality of education, and development of infrastructure and innovation potential, i.e. criteria directly linked with "Industry 4.0". The Association for Industrial Internet, a brainchild of Rostelecom and Roscosmos, can be considered Russia's first step towards the transition to "Industry 4.0". The development must focus on digital design and simulation, new materials, additive technologies, industrial internet and robototronics. At its early stages, the roadmap will cover at least eight industries. Some projects are already underway, e.g. Volgobus unmanned commercial vehicles in the auto industry or the Arctic Project 22220, world's largest nuclear-powered icebreaker, in the ship-building industry. Financial support for high-potential projects will be extended by Vnesheconombank [13].

2.2   Moroccan I4.0 Projects

Not to grow is to slide backwards. To survive in a competitive market, industries should improve their businesses and the quality of their services and products. As an emerging country, Morocco took the first steps towards industry 4.0. After deploying two main sector strategies since 2000 to know "E-Morocco" and "Morocco Numeric 2013", it also adapts an industrial strategy for the long-term up to 2030 which covers a broad scope of topics: smart cities, innovation, education and training, digitalization of the State, electronic commerce and the industry of course [14]. We find evidence for a tendency towards the co-design of policy measures by public and private stakeholders, which, if it persists, could lead the way towards a more effective industrial policy [15] in several sectors, mainly in Agri-Business, Mining, Aerospace, Automotive and Textile industries.

The 6th December 2017, MOROCCO MAGAZINE INDUSTRY, Organizes the first edition of "the morning of the industry", a meeting around: Industry 4.0: Successful Industrial Revolution. More than 300 professionals, decision makers, including the Secretary of State for Investment, Othman El Ferdaous, participated in this meeting. The event was also marked by the presence of eminent scientists such as Jean-Claude Bocquet, a doctor in CAD, or renowned industrialists and leaders such as Fikrat, CEO of Cosumar or Jean Claude Serbes, director of Thales 3D [16].

The same day, OCP Group has announced that it has formed a joint venture (JV) with IBM to provide digital and IT services in sectors ranging from agriculture to industry across Africa. The JV is called TEAL Technology Services and will initially support OCP in accelerating its digital transformation and enhance efficiencies in its business operations. The JV will implement advanced technologies, including analytics, cognitive and internet of things (IoT), and will develop customized services, leveraging OCP and IBM expertise, whilst also looking to generate employment in the IT sector [17].

On the occasion of the inauguration of its Dassault System Learning-Lab, the Higher School of Engineering Sciences and Technologies (ESSTI) of Rabat organized a seminar on March 12, 2018 under the theme "Industry 4.0, ambitions and challenges for Africa".

In addition of being the first laboratory specifically dedicated to Industry 4.0, the ESSTI 3DS lab is also the first academic certification center in Africa on Dassault Systems tools (Dassault is a French Aerospace manufacturer) [18].

Earlier in October 2017, the Moroccan textile industry took a close look at the fourth industrial revolution that is invading the world. On the sidelines of "Maroc in Mode" and "Sourcing Maroc" events, organized by the AMITH, the association of textile professionals discussed the implications of the changes in fashion 4.0 that it sees as a competitive lever for "Fast Fashion" [19].

3 Thing-Based SOA for I4.0

3.1 IoT Related Paradigms

The Internet of Things promises the easy integration of the physical world into computer-based systems. In effect, real-world objects become connected to the virtual world, which allows for the remote sensing of as well as the remote acting upon the physical world by computing systems. Improved efficiency and accuracy are expected from this paradigm shift [20].

Thing/Device. An asset/equipment, system or component that can be tracked. This can include everything from cell phones to coffee makers, washing machines, headphones, lamps, wearable devices and more. A device can also be a component of a machine, such as a jet engine of an airplane or the drill of an oil rig.

Wireless Sensors Network (WSN). Consists of spatially distributed autonomous sensors to monitor physical or environmental conditions, such as temperature, sound, vibration, pressure, motion, or pollutants and to cooperatively pass their data through the network to a main location. Wireless sensors are networked and scalable, consume very little power, are smart and software programmable, capable of fast data acquisition, reliable and accurate over the long term, cost little to purchase and install, and require no real maintenance. They generally consist of a base station (or "gateway") that can communicate with several wireless sensors via a radio link. Data is collected at the wireless sensor node, compressed, and transmitted to the gateway directly or, if required, uses other wireless sensor nodes to forward data to the gateway. The transmitted data is then presented to the system by the gateway connection. Sensor networks are the key to gathering the information needed by smart and ubiquitous environments, whether in buildings, utilities, industrial, home, shipboard, transportation systems, automation, education, or elsewhere [21].

Industrial Wireless Sensor (IWS). Is the application of Wireless Sensor Networks in Industrial field. IWS play a vital role in the Industry 4.0 framework, and can be used for smart factories and intelligent manufacturing systems [22].

3.2 Service-Oriented Architecture

The SOA paradigm is an advance methodology that allows the construction of loose-coupled distributed systems, standards-based, and protocol independent distributed computing. For this reason, the SOA paradigm along with the Web Services are becoming in new way to heterogeneous systems integration [23]. In contrast to EAI, which deals with linking enterprise applications, so they can communicate with one another (by means of an intelligent reasoning engine) and carry out batch data transfers, SOA provides transactional data transfers. For SOA implementation, we need to use an integration platform, called ESB (Enterprise Service Bus). An ESB, viewed as a new generation of EAI, is actually a middleware providing integration facilities built on top of industrial standards such as XML, SOAP, WSDL. A service can be exposed as an API (SOAP or Rest) so that other components can communicate with it.

Traditional SOA involves three main actors that interact directly with one another: a Service Provider, a Service Consumer, and a Registry for services. Any service-oriented middleware adopting this architecture supports three core functionalities: Discovery, Composition of, and Access to services.

- Discovery: is used to publish (register) services in registries that hold service metadata and to look up services that can satisfy a specific request.
- Composition: of services is used when discovered services are unable to individually fulfill the request. In such case, existing services are combined to provide a new convenient functionality. The composed services can further be used for more complex compositions.
- Access: enables interaction with the discovered services.

Core Features.

- **Interoperability**: Enable the communication between heterogeneous systems/applications.
- **Transformation**: The ability of the ESB to convert messages into a format that is usable by the consumer application.
- **Routing**: The ability to determine the appropriate end consumer or consumers based on both pre-configured rules and dynamically created requests.
- **Monitoring**: ESB provides an easy method of monitoring the performance of the system, the flow of messages through the ESB architecture, and a simple means of managing the system in order to deliver its proposed value to an infrastructure.
- **Security**: ESB security involves two main components - making sure the ESB itself handles messages in a fully secure manner, and negotiating between the security assurance systems used by each of the systems that will be integrated.

Benefits.

- Easy to plug-in/plug-out and loosely coupled
- Scalable, Distributable and Reusable
- Cost-effective to ensure high availability and scalability
- Format and business validations and controls
- Reduce siloed systems which require duplication of data and greater ability to expose services to personal devices.

3.3   I4.0 Dimensions and Principles

The attention should be paid to three types of integration: horizontal, vertical and end-to-end integration [1, 3, 24].

- Horizontal integration: refers to a generation of value-creation networks involving integration of different agents such as business partners and clients, and business and cooperation models. It is including the cross-linking of value creation modules along with the material flow along the Supply Chain, from inbound to outbound logistics, passing by Production Management. Value creation modules are defined as the interplay of different value creation factors: Equipment, Human, Organization, Process and Product.

- Vertical networking: concerns smart production systems, e.g.: smart factories, smart products, the networking of smart logistics, production and marketing and services, with a strong needs-oriented. It is including the cross-linking of value creation modules with the different value chain activities (e.g., Procurement, Sales and Marketing…).
- End-to-end integration is targeted at gaining on product design, manufacturing and the customer. It is including the cross-linking of stakeholders; products and equipment along the product life cycle, from raw material acquisition phase and ending with the end-of-life phase.

Below a micro perspective of I4.0 covering the horizontal integration as well as the vertical integration, in addition of the end-to-end engineering dimension [3] (Fig. 1).

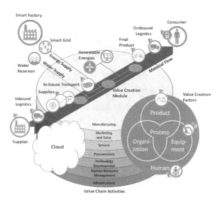

Fig. 1. Micro perspective of Industry 4.0

However, according to a Deloitte study for Switzerland' manufacturing companies [25], there is a forth characteristic which is the "impact of exponential technologies" as an accelerant or catalyst that allows individualized solutions, flexibility and cost savings in industrial processes.

I4.0 already requires automation solutions to be highly cognitive and highly autonomous.

Artificial Intelligence (AI), Advanced Robotics and Sensor Technology have the potential to increase autonomy further still and to speed up individualization and flexibilization.

Under an Industry 4.0 strategy, to upgrade a factory to smart and autonomous there is two design concepts [10]:

- Interoperability: digitalization, communication, standardization, flexibility, real-time, responsibility and customizability.
- Consciousness: Predictive maintenance, decision making, intelligent presentation, self-aware, self-optimization and self-configuration.

By bridging the physical and the virtual worlds, the purpose is not to replicate all human intelligence, it is to enhance and scale human capabilities and expertise.

Hence, IBM, leader in Cognitive Systems, is recommending using the term Augmented Intelligence rather than Artificial Intelligence.

By combining the worker's expertise with the effectiveness of their tools, we can determine if the right worker/tool combo is in place to get work done. And we can monitor environmental conditions and energy consumption in the factory to reduce energy usage.

After, discovering the resulted complexity of these new systems, a new reference architecture is required to ensure scalable systems and solutions with high-availability, security and optimized processes.

4   The Proposed Thing-Based Service-Oriented Architecture for Industry 4.0

Integration remains a pet peeve of I4.0 design and implementation. This specific and complex context is imposing several challenges:

- Ultra-large number of Things, generating a large-scale data
- Real time operations visibility anywhere and anytime
- Deep functional and technical heterogeneity
- Access & Security
- Mobile/remote process and operations console
- Monitoring & Control, especially with virtual expert and remote diagnostics

The schema below is presenting the proposed architecture to Integration of things and services (Fig. 2).

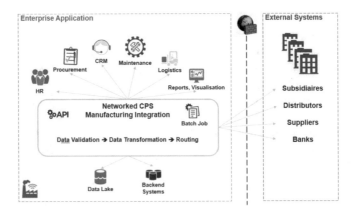

Fig. 2. Networked CPS manufacturing integration architecture

The Middleware Layer is hosting multiple web services and batches jobs enabling the communication between the different enterprise, sites and external applications and systems.

Middleware Layer will carry business services and processes as part of the IS transformation to a service-oriented architecture. It ensures the publication and assembly of business services, as well as the orchestration of processes via a Business Process Management (BPM) Platform (Fig. 3).

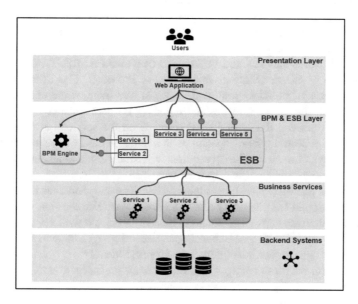

Fig. 3. Integration layers

The synchronous interfaces are based on web services exposed to be called by the consumer applications to retrieve data from the backend systems. Backend systems are dealing with databases and data processing components. Data sources can be managed under a new approach called Data Lake. Data lake is different of Data Warehouses. It is a repository of data stored in its natural format. Different format types:

- Structured (rows and columns): from line of business applications
- Semi-structured (CSV, logs, XML, JSON)
- Unstructured (emails, documents, PDFs): mobile apps, IoT devices and social media.
- Binary multimedia format (images, audio, video).

It contains all enterprise data including raw copies of source system data and transformed data used for integration, reporting, visualization, analytics and machine learning.

The web services are set up to support the Functional Services. The web service is characterized by a standardization of implementations, a remote location of services and a recovery of the access interface allowing the execution of the corresponding processing. For this reason, the web service is most important component in of SOA. The implementation is carried by an ESB platform.

ESB could perform many types of control and validation (pattern, syntax, functional…) on input data. The platform is able to accept messages sent in all major protocols, and convert them to the format required by the end consumer.

This Integration platform will manage internal applications communications such as HR, Financial, Procurement, Logistics and Maintenance…, in addition of external flows with subsidiaries, distributors, suppliers, banks… This configuration comes along with infrastructure architecture design to take up system heterogeneity and manage the different exchange protocols.

The Synchronous services will be exposed as web APIs, REST (Representational State Transfer) or SOAP (Simple Object Access Protocol). The ESB platform will enforce the usage of policies, control the access, and monitor the activity with collecting and analyzing usage statistics and reporting on performance.

The choice between REST or SOAP is based on the service logic, security constraints and the expected type of operations to handle on data. However, Restful APIs are easier to use and more flexible. It requires less bandwidth and resource and permits different data format such as Plain test, HTML, XML, JSON…

The services will be exposed as endpoint Alias. The endpoint alias information will override the endpoint URL to encapsulate the original service location/path and give a representative alternate name. The endpoint alias is associated with a specific Transport Protocol HTTP or HTTPS. For enhancing the security, HTTPS will be used for all exposed APIs to ensure confidentiality, integrity and identity.

ESB platform handles messages in a fully secure manner and provide groups and users management using ACL (Access Control Lists) to route the messages to the appropriate end consumers.

Integration platforms could implement also asynchronous interface. It will ensure Job scheduling, Messaging Queues management, and secure file transfer and delivery for internal and external systems along with ESB tasks of Data Validation, Transformation, Routing, and Monitoring services.

The most important feature of the Integration platform, is enabling an easy method of monitoring the performance, the flow of messages through the ESB. The service users, end consumers, should get their data when it is demanded, under the agreed response time and respecting the expected requirement and quality. Generally, companies impose what we call Service-level agreement (SLA) as a commitment between the service provider and the service consumer/client. They agree on the expected quality, availability and responsibilities depending on the service Criticality and Priority.

5 Conclusion and Perspectives

Industry 4.0 concepts are perceived as highly complex because upgrading smart manufacturing requires new digital strategies and processes and the use of IoT devices supporting automatic and remote management. The development of these approaches imposes the need of a reference architecture offering strong capabilities of scalability, security and optimized business processes.

Firstly in our research we aimed to discover the multiple Industry 4.0 projects started by the greatest manufacturing countries, in addition of the Moroccan initiatives to follow this industrial revolution. Then we have presented different I4.0 paradigms to simplify these new approaches for academic or professional purposes. After, discovering the resulted complexity of the new systems, a new reference architecture is required to ensure scalable systems and solutions with high-availability, security and optimized processes. Finally, we proposed a thing-based architecture, SOA friendly responding to the industry 4.0 challenges regarding high availability in addition of collecting, integrating and processing real time data.

References

1. Recommendations for implementing the strategic initiative INDUSTRIE 4.0, Final report of the Industrie 4.0 Working Group, acatech (2013)
2. Germany: Industrie 4.0, Digital Transformation Monitor, European commission (2017)
3. Stock, T., Seliger, G.: Opportunities of sustainable manufacturing in industry 4.0. Procedia CIRP **40**, 536–541 (2016)
4. European Commission: Factories of the Future (2015). https://ec.europa.eu/digital-single-market/en/blog/factories-40-future-european-manufacturing. Accessed Mar 2018
5. https://www.techradar.com/news/what-is-industry-40-everything-you-need-to-know. Accessed Mar 2018
6. https://www.process-worldwide.com/usa-industry-40-the-american-way-a-536602/. Accessed Mar 2018
7. https://www.entreprises.gouv.fr/innovation-2030/home?language=en-gb. Accessed Mar 2018
8. Rethink manufacturing, Designing a UK industrial strategy for the age of Industry 4.0, KPMG (2017)
9. Center for Strategic International Studies. https://www.csis.org/analysis/made-china-2025. Accessed 1 Oct 2018
10. Qin, J., Liu, Y., Grosvenor, R · A categorical framework of manufacturing for industry 4.0 and beyond. Procedia CIRP **52**, 173–178 (2016)
11. KPMG, Internet of Things: An Indian context (2016)
12. Japan. http://www.meti.go.jp/english/press/2016/0428_04.html. Accessed Mar 2018
13. Russian News Agency. http://tass.com/sp/948066. Accessed Mar 2018
14. Bakkari, M., Khatory, A.: Industry 4.0: Strategy for More Sustainable Industrial Development in SMEs (2017)
15. Vidican-Auktor, G., Hahn, T.: The effectiveness of Morocco's Industrial Policy in Promoting a National Automotive Industry. Discussion Paper/Deutsches Institut für Entwicklungspolitik (2017). ISSN 1860-0441
16. https://www.leconomiste.com/article/1021112-l-industrie-4-0-aux-portes-du-maroc. Accessed Mar 2018
17. https://www.worldfertilizer.com/project-news/06122017/ocp-group-and-ibm-form-jv/. Accessed Mar 2018
18. http://www.leseco.ma/enseignement/535-enseignement/64445-l-essti-aide-le-maroc-a-entrer-dans-la-4e-revolution-industrielle.html. Accessed 1 Oct 2018

19. http://www.amith.ma/Portail/pdf/Communiqu_de_presse_n4_Industrie_4_0_pour_le_ textile.pdf. Accessed Mar 2018
20. Issarny, V., Bouloukakis, G., Georgantas, N., Billet, B.: Revisiting Service-oriented Architecture for the IoT: A Middleware Perspective, HAL Id: hal-01358399 (2016). https:// hal.inria.fr/hal-01358399. Accessed Mar 2018
21. El Kho, F., Tahiri, A., Aknin, N., El Kadiri, K.: A novel e-learning model management system based on business intelligence tools and WSN technology. IJERED **2**, 1–6 (2014). ISSN 2320-8708
22. Li, X., Li, D., Wan, J., Vasilakos, A.V., Lai, C.-F., Wang, S.: A review of industrial wireless networks in the context of Industry 4.0. Wirel. Netw. **23**(1), 23–41 (2017)
23. Herrera-Quintero, L., Maciá-Pérez, F., Marcos-Jorquera, D., Gilart-Iglesias, V.: SOA-based Model for the IT integration into the intelligent transportation systems. In: IEEE ITSC 2010 Workshop on Emergent Cooperative Technologies in Intelligent Transportation Systems (2010)
24. Mrugalska, B., Wyrwicka, M.K.: Towards lean production in industry 4.0. Procedia Eng. **182**, 466–473 (2017)
25. Schlaepfer, R.C., Koch, M., Merkhofer, P.: Industry 4.0: challenges and solutions for the digital transformation and use of exponential technologies. The Creative Studio at Deloitte, Zurich (2014)

Ubiquitous Platform as a Service for Large-Scale Ubiquitous Applications Cloud-Based

Marwa Zaryouli[(⊠)] and Mostafa Ezziyani

LMA Laboratory, Faculty of Sciences and Techniques of Tangier, Abdelmalek
Essaâdi University, Tangier, Morocco
zaryouli.marwa@gmail.com

Abstract. Ubiquitous computing have many challenges including the limitation
of multi-domain, intensive mobility of users, the lack of a uniform namespace
limited resources, lack of scalability, intensive applications... all of this leads
researchers to provide ubiquitous hybrid architectures that are typically cloud-
based. In this article, we propose a cloud-based ubiquitous application platform,
Leveraging the potential of cloud computing in supporting large-scale ubiquitous
computing applications.

The principal goal of this architecture is to remove the limitations on pure
ubiquitous architectures. Indeed, two categories of services are presented in this
architecture, the first one based on the distance between requesting and
responding entities, and the second one based on the nature of services, both of
which are independent of the type of application.

Keywords: Ubiquitous computing · Cloud-based · Services · Pervasive
computing · Ubiquitous cloud computing

1 Introduction

In recent years, new business and research opportunities have been increasingly
emerging from the need for large scale ubiquitous computing systems (Weiser, 1991),
[5], which is, described as a sort of distributed computing. Including for example
pervasive healthcare, smart cities, and so on. From a system's point of view, they are
characterized by employing large numbers of mobile devices and specialized sensors
connected in a volatile network setup.

In fact, ubiquitous computing is not solely the control of the peripheral environment; rather, a more abroad concept is suggested. They generate and process massive
amounts of real-time data while coping with fluctuation of user demands. However,
after this person enters the building, if he is identified, his profile can be examined and
lighting adjusted based on personal preferences (service based on implicit demand), it
can then be suggested that a ubiquitous computing based environment is in play. This
simple example demonstrates the scenario of a single-domain ubiquitous environment,
by which it can be inferred that processing in the environment is the main basis of
ubiquitous computing.

So ubiquitous computing is divided into single-domain and multi-domain ubiquitous computing [6, 7]: Single-domain indicates that changes in the environment are

© Springer Nature Switzerland AG 2020
M. Ezziyani (Ed.): AI2SD 2019, LNNS 92, pp. 301–310, 2020.
https://doi.org/10.1007/978-3-030-33103-0_30

only applied based on the local knowledge of the environment (user behavior and characteristics in a local situation) and not on previous information or new data derived from other domains. In contrast, multi-domain ubiquitous computing refers to the exploitation of several ubiquitous contexts, which interact with each other; the information derived from each of these is applicable in other domains. To give an example to a multi-domain system we have a user who wants leave his office (domain 1) and go home (domain2), On the way, he visits a medical clinic because of a headache (domain 3) and the physician prescribes some medicine based on the examination and the user's physical condition at work.

The user then visits a drugstore (domain 4), picks up the medicine, and eventually returns home.

Based on the type of processing core, ubiquitous computing systems are divided into pure and hybrid computing. In contrast, hybrid systems are those in which ubiquitous computing is integrated with another type of powerful computing method [8], mostly cloud computing so that cloud resources and services can also be utilized along with ubiquitous resources and services. The concept of hybrid cloud-based systems has been utilized in other types of application domains, such as cyber-physical systems [9] and manufacturing systems [10]. The main reason for introducing hybrid systems is to explain the deficiencies and problems of pure systems.

This paper proposes to address the challenges in an environment large-scale ubiquitous computing systems by consolidating the core concepts of two computing paradigms—Cloud computing and Ubiquitous computing. Conceptually, ubiquitous computing implies using massive numbers of small computational entities to reach all aspects of human life.

The main reason for introducing hybrid systems is to explain the deficiencies and problems of pure systems. These are as follows: Limitation of being multi-domain, Lack of uniform namespace in the ubiquitous system, Incapability of users' ubiquitous mobility [11], Restrictions of resources and lack of scalability of a pure ubiquitous system, Intensive applications [12].

According to these limitation issues, access to an ideal ubiquitous computing system is not possible individually via access to local resources and services. To eliminate the deficiencies of pure ubiquitous systems, one should consider regulation of hybrid systems for offloading some of the requests to powerful computing, such as cloud computing.

The paper is organized as follows: Sect. 2 presents other researchers about hybrid architectures and examines their services and categorization in various ubiquitous environments. Section 3 explains the reason for focusing on ubiquitous cloud computing. Section 4 while discussing the general features of an ideal architecture for ubiquitous cloud computing and examining its interactions and the applicability of the services in different ubiquitous environments. Finally, Sect. 5 presents the conclusion.

2 Related Work

There is a lot of research in the context of hybrid ubiquitous cloud architectures. In [2], due to the limitations of mobile devices as well as network shortcomings, an architecture is presented based on location-based services. By dividing tasks and utilizing outsourcing techniques, these services process requests that require intensive resources within the cloud and those, which are lightweight on mobile devices. In this architecture, all communications between users and the cloud are end-to-end and through internet. Hence, when the network is disconnected, users have limited or no access to their intended services, which affects real-time applications.

In addition, in the description of this architecture, there is no discussion on how to control the environment and provide local services.

In [1], a platform named EUPaaS (Elastic Ubiquitous Platform as a Service) is proposed. Its main purpose is to provide real-time data services to a wide range of ubiquitous applications by utilizing the ability of real-time data processing in the cloud. Although this platform can be used in diverse applications, including the medical field, it must first be noted that its details and subsystems functionality has not yet been well defined. Secondly, the EUPaaS layer communications with the ubiquitous and cloud structures are not clear.

In [3] addresses ambient aided living systems and introduces a context-aware architecture named as COCAMAAL, which is based on five different clouds. In addition to managing sensor data and extracting context information, the architecture can monitor user activities, search for, and provide appropriate situational services for them. As in most hybrid architectures, all processes are performed on clouds. In addition, there is a local station for each user domain that is only responsible for collecting sensor data and sending them to the cloud. These stations play no role in providing local services to users since this would only increase the costs of deployment.

In [4], the UbiCloud hybrid architecture is for intensive applications, such as image processing, data mining, multi-media applications, etc., in which all services are assumed identical in terms of diversity and access type. In this architecture, regardless of failing to introduce architectural subsystems, there was also no discussion on the details of ubiquitous and cloud structures. In addition, since context information is not collected, users cannot influence their surrounding environment nor control it (Table 1).

3 Comparison Between Ubiquitous Computing, Cloud Computing and Ubiquitous Cloud Computing

For ubiquitous computing to be desirable, it must respond to user needs based on existing technologies. A desirable ubiquitous computing system must be capable of eliminating the pure system's deficiencies discussed in the first section of the current paper. Moreover, it should be able to process any kind of request with any number of required resources.

Table 1. Characteristics of some hybrid ubiquitous computing architectures

Name	Type of resources	Type of services	Other Limitations
[2] No name	Lightweight intensive	Cloud-based context-aware	Limited or no access to services, when internet is disconnected
[1] EUPaaS	Intensive	Cloud-based	The lack of architecture details and subsystems functionality; no access to services, when internet is disconnected
[3] COCAMAAL	Intensive	Cloud-based	No access to services, when internet is disconnected; increasing the costs of deployment due to utilizing local stations
[4] UbiCloud	Intensive	Cloud-based	The lack of architecture details; no access to services, when internet is disconnected

Now, the questions to ask are what kind of computing can provide this ideal, if software and hardware augmentation of pure ubiquitous computing are able to eliminate these challenges, and if using another computing system will provide the required capabilities on its own. Ubiquitous computing does not claim to have all the necessary capabilities. Furthermore, a stronger computing system, like cloud computing, is not be able to fulfill ubiquitous demands for several reasons, such as a lack of context knowledge, the impossibility of controlling local devices and equipment, the limitation in the type of calling for services, and so on. Therefore, the best solution is a kind of hybrid computing which can provide the required capabilities by combining strong computing with pure ubiquitous computing.

In this regard, research conducted in recent years has focused on two types of hybrid computing: ubiquitous cloud computing [13].

Ubiquitous Cloud Computing is a type of multi-domain hybrid ubiquitous computing, which employs elastic and unlimited cloud resources as the head layer and context-aware ubiquitous technologies as the bottom layer to provide a vast number of services for anyone, anything, anywhere, and anytime, in various types and quantities regardless of heterogeneity in devices and equipment. Moreover, users should not be forced to pay for using some kind of services.

Ubiquitous cloud computing is a combination of cloud and ubiquitous technologies, both of which together can respond to a vast variety of user requests. An architecture based on this kind of computing should be capable of providing different services with any kind or volume of resources for any users or things (actuator devices and user equipment) found in different areas of the ubiquitous environment regardless of the request time.

This table compares ubiquitous computing and ubiquitous cloud computing against the features expected from ideal ubiquitous computing (Table 2).

Table 2. Comparison between ubiquitous computing, cloud computing and ubiquitous cloud computing

Expected features	Ubiquitous computing	Cloud computing	Ubiquitous cloud computing
Type of services	Lightweight	Various types	Various types
Heterogeneity	Just in ubiquitous components	Just in cloud Components	Ubiquitous, cloud components
Covered domain	Limited	Unlimited	Unlimited
Mobility of users	Based on the connection type between domains	Based on the connection type between domains	Based on the connection type between domains
Calling for services	Explicit, Implicit	Explicit	Explicit, Implicit

Type of Services
In ubiquitous cloud computing, any type of service can be provided because of the use of clouds.

Heterogeneity
In cloud computing, this heterogeneity exists in cloud components. However, in ubiquitous cloud computing, ubiquitous, mobile, and cloud components can be non-uniform.

Covered Domain
In cloud computing, if specified in the service level agreement, the user can consistently access all cloud services and capabilities from any location. In ubiquitous cloud computing, users must be able to consistently benefit from ubiquitous services and capabilities in every condition, even if these services are limited.

Mobility of Users
The users of ubiquitous cloud computing should be able to access all the provided services regardless of geographic location. In ubiquitous computing, if the user is geographic location changes, the use of ubiquitous services will still be possible. However, users are only able to benefit from ubiquitous services in the domains related to their current domain [12].

Calling for Services
In ubiquitous computing, services can be explicitly and implicitly called due to the presence of numerous.

Peripheral Sensors. However, in cloud computing, services are only called explicitly and upon direct request by the user.

4 The General Features of an Ideal Architecture for Ubiquitous Cloud Computing

4.1 Architecture for Ubiquitous Cloud Computing

This demonstrates a connection between ubiquitous domains and the computing cloud by considering the basic architecture of each.

This multi-domain hybrid architecture is constituted of three layers: the user layer (the lowest layer), ubiquitous layer (middle layer), and cloud layer (highest layer). The relation between these layers is hierarchical [14] and the basis of connection is the ubiquitous layer (equal to the distributor layer in [14] 's definition of layers).

In Fig. 1, n numbers of ubiquitous domains are present in the ubiquitous layer, each of which include a number of users in the user layer and all of which are connected through a cloud in the cloud layer. Each ubiquitous domain is capable of connecting to other components in three methods. In the first simplest method, to receive user requests and send the proper output after their processing, a connection is established with users through the current domain's local devices and equipment. In this case, the user requests are in need of the local domain's services.

There are a variety of devices and equipment, which play the role of a link connecting users to the system.

Sensor Devices
Such as smoke sensors, body sensors or motion detection sensors, which enter inputs into the system after being installed in the environment.

Smart Equipment
Such as smart mobile phones, pocket computers or tablets with the capability of both entering inputs and displaying results to users.

Non-smart Equipment
Such as the keyboard, control panel or light keys, which are solely capable of entering inputs into the system.

Actuator Devices
Such as lamps, monitors or blood pressure alarms which solely show the resulting outputs.

Explicit Method
In this method, inputs are entered into the system directly by the users and through the common input tools, such as the keyboard, mouse, touchscreen, microphone, camera, light keys, and controlling keys on a panel. Explicit inputs are generally entered into the system through both smart and non-smart equipment.

Implicit Method
In this method, inputs are entered into the system accidentally through environmental sensors, such as smoke, temperature, pressure, weight, motion, connection, light, and the other sensors. Although implicit inputs are usually entered into the system through environmental sensors.

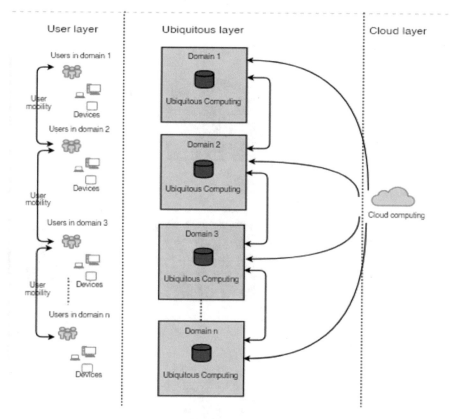

Fig. 1. Architecture layers based on the proposed definition for ubiquitous cloud computing

In a ubiquitous system, one can attribute several features to inputs, whether implicit or explicit. These features greatly affect the functioning of a ubiquitous system.

In Fig. 2, we present an example of the interaction between users in domain 1 and the system for sending input requests and receiving output results. In this figure, user requests are first sent to the ubiquitous management system in domain k. This system will then decide to process the requests locally or send them to the cloud or one of the adjacent domains for processing.

In this architecture, users of a domain can make connections to other close by users of the same domain via their own equipment.

4.2   Nature of Services

Data Services
Group of various data and information, which can be utilized in the form of a service, directly by the user in different applications or by other types of services.

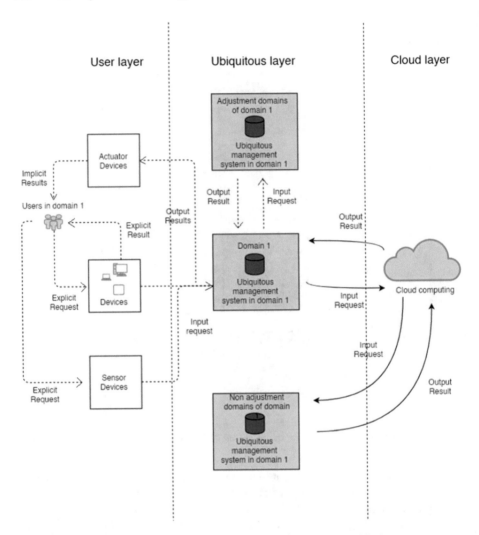

Fig. 2. Interaction between users of a ubiquitous domain and the ubiquitous management system for sending requests and receiving results.

Software Services
Software services are a collection of software and applications, which can be employed for various functions, either directly by the user or by other user requested services.

Hardware Services
A hardware service is the usage of a group of hardware, resources, such as CPU, RAM, storage, sensors, and so on. These services can be identified and utilized in various applications. Hardware services are usually used by other ubiquitous services, especially software services.

Context-Aware Services

Context-aware services are all control actions and capabilities that can be directly applied by the user or by other services requested by the user into the context. Some of these services may include enabling or disabling of a series of peripheral equipment.

Near Services

If a number of services are provided by the existing devices and equipment in the user layer and these can be provided solely for close by users or the services requested by nearby users, then these services are called near services.

Local Services

In a ubiquitous environment, if a number of services are provided by the ubiquitous system's processing core in a specific domain and only for users of that domain, these services are called local services.

Global Services

In a hybrid ubiquitous environment, global services can be employed at all places by users or their requested services. Generally, these services cannot be provided without the usage of clouds.

Far Services

In a hybrid ubiquitous environment, far services refer to the services supplied by the processing core of each non-adjacent domain of the current domain; far services can be provided by a cloud for current domain users or other services requested by them.

5 Conclusion

A ubiquitous cloud architecture was introduced as basic architecture. Based on the requirements expected of desirable ubiquitous computing and the structure of the basic architecture, we proposed for this architecture: one based on the nature of services and the other on the distance between the service requester and the service provider.

From the perspective of ubiquitous services applicability, these services must be defined and categorized in such a way so as to allow their use in any application. In addition, due to the importance of security in all computational architectures, the most critical security concerns related to ubiquitous cloud computing architectures

References

1. Li, F., Dustdar, S., Bardram, J., Serrano, M., Hauswirth, M., Andrikopoulos, V., Leymann, F.: EUPaaS-elastic ubiquitous platform as a service for large-scale ubiquitous applications
2. Choi, M., Park, J., Jeong, Y.S.: Mobile cloud computing framework for a pervasive and ubiquitous environment. J. Supercomput. **64**(2), 331–356 (2013). https://doi.org/10.1007/s11227-011-0681-6
3. Forkan, A., Khalil, I., Tari, Z.: CoCaMAAL: a cloudoriented context-aware middleware in ambient assisted living. Future Gener. Comput. Syst. **35**, 114–127 (2014). https://doi.org/10.1016/j.future.2013.07.009

4. Youssef, A.: Towards pervasive computing environments with cloud services. Int. J. Ad Hoc Sens. Ubiquitous Comput. **4**(3), 01 (2013). https://doi.org/10.5121/ijasuc.2013.4301

5. Weiser, M.: The computer for the 21st century. Mobile Comput. Commun. Rev. **3**(3), 3–11 (1999). https://doi.org/10.1145/329124.329126

6. da Rocha, R.C.A., Endler, M.: Domain-based context management for dynamic and evolutionary environments. In: 4th Middleware Doctoral Symposium, p. 16. ACM (2007)

7. Dey, A.K., Abowd, G.D., Salber, D.: A conceptual framework and a toolkit for supporting the rapid prototyping of context-aware applications. Hum. Comput. Interact. **16**(2), 97–166 (2001). https://doi.org/10.1207/S15327051HCI16234_02

8. Hingne, V., Joshi, A., Finin, T., Kargupta, H., Houstis, E.: Towards a pervasive grid. In: International Parallel and Distributed Processing Symposium, p. 8. IEEE (2003)

9. Mourtzis, D., Vlachou, E.: Cloud-based cyber-physical systems and quality of services. TQM J. **28**(5), 704–733 (2016). https://doi.org/10.1108/TQM-10-2015-0133

10. Xu, X.: From cloud computing to cloud manufacturing. Robot. Comput. Integr. Manuf. **28**(1), 75–86 (2012). https://doi.org/10.1016/j.rcim.2011.07.002

11. Al-Rwais, S., Al-Muhtadi, J.: A context-aware access control model for pervasive environments. IETE Tech. Rev. **27**(5), 371–379 (2010). https://doi.org/10.4103/0256-4602.63847

12. Al Ali, R., Gerostathopoulos, I., Gonzalez-Herrera, I., Juan- Verdejo, A., Kit, M., Surajbali, B.: An architecture-based approach for compute-intensive pervasive systems in dynamic environments. In: 2nd International Workshop on Hot Topics in Cloud Service Scalability, p. 3. ACM (2014)

13. Van der Merwe, J., Ramakrishnan, K.K., Fairchild, M., Flavel, A., Houle, J., Lagar-Cavilla, H.A., Mulligan, J.: Towards a ubiquitous cloud computing infrastructure. In: 17th IEEE Workshop on Local and Metropolitan Area Networks (LANMAN), pp. 1–6. IEEE (2010)

14. High Availability Campus Network Design - Routed Access Layer using EIGRP, Cisco Validated Design II. http://www.cisco.com/application/pdf/en/us/guest/netsol/ns431/c649/ccmigration_09186a00808f6c34.pdf

Author Index

M. Ezziyyani (Ed.): AI2SD 2019, LNNS 92, pp. 311–312, 2020.
https://doi.org/10.1007/978-3-030-33103-0

Printed in the United States
By Bookmasters